PARENTING ON EARTH

PARENTING ON EARTH

A Philosopher's Guide to Doing Right by Your Kids—and Everyone Else

ELIZABETH CRIPPS

The MIT Press
Cambridge, Massachusetts
London, England

The MIT Press would like to thank the anonymous peer reviewers who provided comments on drafts of this book. The generous work of academic experts is essential for establishing the authority and quality of our publications. We acknowledge with gratitude the contributions of these otherwise uncredited readers.

This book was set in Adobe Garamond and Berthold Akzidenz Grotesk by Jen Jackowitz. Printed and bound in the United States of America.

Library of Congress Cataloging-in-Publication Data

Title: Parenting on earth : a philosopher's guide to doing right by your kids and everyone else / Elizabeth Cripps.
Description: Cambridge, Massachusetts : The MIT Press, 2023. | Includes bibliographical references and index.
Identifiers: LCCN 2022033283 (print) | LCCN 2022033284 (ebook) | ISBN 9780262047586 (hardcover) | ISBN 9780262372800 (epub) | ISBN 9780262372817 (pdf)
Subjects: LCSH: Parenting. | World citizenship. | Globalization—Social aspects. | Environmental ethics. | Environmental responsibility.
Classification: LCC HQ755.8 .C736 2023 (print) | LCC HQ755.8 (ebook) | DDC 649/.1—dc23/eng/20220725
LC record available at https://lccn.loc.gov/2022033283
LC ebook record available at https://lccn.loc.gov/2022033284

10 9 8 7 6 5 4 3 2 1

For my beloved daughters

Contents

CONCLUSION *195*

Note to Readers

Parenting on Earth is for parents, biological or adoptive. It is for stepparents and foster parents, or those thinking of having kids. But neither our love nor our hope is limited to our own children. I hope it will also interest aunts, uncles, grandparents, godparents, family friends, and those involved in communal child-rearing arrangements.

In planning this book, I have been finding my own path; in writing it, I speak primarily to those who are also comfortably off, from the global perspective. That includes most parents in rich countries like mine, though not only or all of them. That is not to deny agency to families already coping with searing poverty, rampant disease, or a devastated climate. Quite the reverse: global decision-makers should listen to precisely these voices. But it is those in cultures of material affluence and abundant waste who need to change. For convenience, I will often use "we," for such parents, but of course "we" still experience varied challenges, or different degrees of privilege. I hope that, too, will be apparent.

In case you want a reminder, I've summarized the core ideas of each chapter before moving on to the next one. I've also included copious endnotes for anyone who wants to probe further into my sources or follow me on academic tangents. If that's not you, you can read this almost as fully (and much more easily!) without them.

INTRODUCTION

September 20, 2019. It's the Global Climate Strike and I have two six-year-olds in tow. We march among thousands through Edinburgh's Old Town, waving handmade placards, joining in the singing, sharing the cry of "Climate Justice Now." The United States is nearly three years into the Trump administration. The UK government talks the talk on climate change, then approves a coal mine. It's hard to keep positive. But this crowd, buoyed with determination, helps me do just that. As for my daughter and her friend, on this ominously hot day, they forget politics and dance in the water features outside the Scottish Parliament.

Three months later, it's Christmas Day and I'm on Ballycastle Beach in Northern Ireland. My daughters and nephews race across sand, dipping wellies into waves. I watch them with my dad and my sister, wrapped in winter layers. We have all loved this beach, child to adult. We've run along it, cried on it, swum in water so cold it's like a rite of passage. My brother-in-law proposed on it. My dad first came here, a toddler, in 1951, my youngest daughter in 2015 when she was six weeks old. We could all close our eyes anywhere in the world and see its landmarks clear as on a high-res photo. I ask myself whether it will be here when my daughters are my age. I know it won't be the same.

I know, too, that my anticipated grief, however real, is comparatively trivial. Families in Bangladesh, in Domenica, have seen their whole lives swept under water.

Another year on, January 2021, we're ten months into a pandemic, and UK schools are closed for the second time. More than two million people have died of COVID-19. Health services are on their knees. I'm homeschooling, running between rooms, checking my seven-year-old's spellings one minute, coaxing my five-year-old to identify *sh* words the next. Once again, we've had to tell our girls that they can't see their grandparents, that holidays are canceled, friendships must survive or wither online. And we're the lucky ones. We haven't had to tell them they'll never see their grandparents again. We haven't needed food banks to keep them alive.

This is the reality of raising kids today.

For me, as for many others, the parent-child relationship is like no other: transformative, intimate, infinitely frustrating, exhausting, emotionally draining, and incredible. Parenting is about love, permeated by love, made terrifying and wonderful by love. It's about experiencing life vicariously through a child's eyes. It's about reading to my daughters, living their joy in a new world shared. It's about the expression in her dad's eyes when my six-year-old babbles with joy because she saw a kingfisher. It's about hope and fulfillment, worry and guilt: investing almost all of oneself in another's well-being.

Right now, parenting involves optimism, because of the many things that *have* got better, over the decades and centuries: the benefits that art and culture, science and innovation (particularly medical science) have given to our children. It also involves fear, on a scale so big we try not to think about it. This book will face up to that fear, and to what it means.

I'm a moral philosopher, and this is a different kind of parenting manual.

For nearly two decades, I've researched collective responsibility, climate justice, and what we as societies and individuals owe to future generations. Teaching this subject, writing papers on it, and discussing it with my students crystallized something for me. The way we live is out of joint with even basic morality: the least controversial notion of what we should do for other people. Then I became a mother, and my investment in the future was suddenly, searingly personal.

Since then, I've been trying to understand what it means for someone like me, living a relatively comfortable life, to be a good parent in this messed-up world. What do we owe our own children, and everyone else? If you're reading this book, it's because you want to know too.

So here's what I think. In countries like the United States or the UK, or anywhere in the so-called Global North, we've got things appallingly wrong. In focusing all our time and money and resources on our own kids—their immediate well-being, their material needs and economic prospects—we've not only failed everyone else, but let *them* down, catastrophically, at the collective level. We also have an incredible opportunity, right now, to get it right.

To show this, I'll start with some ideas that form an intuitive, commonsense approach to morality, and make sense across various philosophical theories. I'll build on them with careful reasoning and critical reflection (the philosopher's stock in trade) to get at the moral basics of being a "good parent" on the one hand, and a "good global citizen," on the other. I'll fit these together to build a manifesto for earned hope in these frightening times.[1]

This book will erase some dilemmas you think you face, but it will make others more real. I'll make a case for changing the way we raise our kids, and how we live with them. Most of all, I'll explain why parents like me should also be social, political, and environmental activists. But even when we've reinterpreted "good parenting," there are still hard choices and psychological hurdles aplenty. I'll be honest about these. I'll lay out the philosophical tools, and some psychological ones, to help you find your own way.

I'll write as the philosopher I am now and the journalist I once was, drawing on the insights of other parents, activists, and experts from other fields. I'm also a mother: muddling through, loving my kids, worrying for them, learning from them even as I try to guide them. I'll write as her too.

A FRAGILE INHERITANCE

As I type this, in May 2022, Russia is two months into its brutal invasion of Ukraine. Alongside the outrage and pity, a specter rises in my mind and those of the people around me: the possibility that one power-crazed man will destroy us all, with the weapons science and geopolitics have given him. My six-year-old asks me, "Who will win?" I look into her beautiful eyes and recall images, horrible and unbidden, that have lain dormant since I was a schoolgirl reading Nevil Shute's post-apocalyptic novel *On the Beach*.[2]

To earn hope, we must first face reality.

I tell my daughter I don't know who will win, but I hope it will be Ukraine. I don't tell her that we could all lose, terribly, or that the nuclear threat is just one more poisoned cherry on the cake. But it is, and I must acknowledge that. What's more, the other global crises facing her generation spring not from a delusional dictator's obsession but, entirely predictably, from the way we live now.[3]

In August 2021, I'm reading the latest report from the Inter-governmental Panel on Climate Change while my children are at school. These reports have been coming out since 1990, and each time they are more decisive. This one says it is "unequivocal" that human influence has warmed the climate. I already know what that means: disease and heatstroke, drowning and fire, drought and starvation. Ten days later, Henrietta Fore, executive director of the children's charity UNICEF, calls the climate crisis "unimaginably dire."[4]

That same month, I have a video call with New York-based educator Harriet Shugarman, founder of the parent activist network Climate Mama, professor of global climate change policy and world sustainability, and the mother of young adults. She tells me that people ask her: "Where can I move?" Her answer? "You can't. If my children aren't safe, nobody's safe."[5]

She's right.

It will take emissions cuts of 45 percent from 2010 levels by 2030 to keep global temperatures from rising to more than 1.5°C above preindustrial levels. In 2030, my girls will turn fifteen and seventeen. Unless something changes, they won't even be old enough to vote. With 2°C of global warming, heat extremes that previously happened twice a century could strike every three years. At 4°C, they could be near-yearly events. According to UNICEF, almost every child on earth faces at least one climate or environmental hazard. Between 2030 and 2050, climate change will kill some quarter of a million people a year.[6]

No wonder, then, that young people are angry and afraid. Of ten thousand aged sixteen to twenty-five, in ten countries, three-quarters are frightened for the future, 56 percent think "humanity is doomed," 65 percent believe governments have failed young people, and 58 percent feel that their generation has been betrayed. And climate change is not the only global hazard our children will face. It's not even the only environmental one.[7]

"It was like going to war every day and this is not an exaggeration." "I have never experienced, and never would want to experience this, in a lifetime. It is unbelievable."

The first quote is from Dr. Pruthu Narendra Dhekane, a Bengaluru doctor. The second is from UK healthcare assistant Michelle Piper. Both were serving on COVID-19 wards. The pandemic caused 14.9 million excess deaths in 2020 and 2021, according to World Health Organization figures. Horrible, lung-bursting, lonely deaths. And this, too, was not only predictable, but predicted.[8]

In September 2021, I speak to Sophie Harman, global health politics professor at London's Queen Mary University. She's just a face on a screen, like everyone else in these pandemic days, but she makes herself abundantly clear. "It was always a when," she tells me. "Not an if." A year earlier, Larry Brilliant, leading epidemiologist, told *The Economist*: "Probably 30 or 50 or all the infectious-disease epidemiologists in the world predicted this almost precisely."[9]

For most parents, there is one mercy: children are less vulnerable to this hideous disease. But youth will not always protect them; nor can we expect such leniency from future pandemics. And, of course, COVID-19 has hurt children *indirectly*, very badly indeed. Jobs lost, lives ruined, schools closed, teachers thrown unprepared into online learning. In the UK, 68 percent of young people aged thirteen to twenty-four said their mental health got worse in lockdown. For those who had already had mental health difficulties, the figure was 75 percent. In another study, parents reported increased behavioral and attentional difficulties.[10]

I want my children to have a flourishing future, not one of lockdowns and fear.

I also want them to have all the advantages of modern medicine. But I can't take even that for granted.

In August 2016, a seventy-year-old was admitted to hospital in Reno, Nevada, with what turned out to be a strain of *Klebsiella pneumoniae*. She died weeks later of septic shock. The Centers for Disease Control and Prevention confirmed that all twenty-six antimicrobials available in the United States had been tried, in vain. Her death made headlines, but it was no isolated incident.[11]

Pneumonia, E. coli, sepsis, UTIs, STDs, bloodstream infections. It's an unpleasant list, but one we should worry about, since they are all developing drug-resistant strains. *Thirty-five thousand* people die every year in the United States of infections resistant to antibiotics (the type of antimicrobial that targets bacteria). In Europe, in 2015, the death toll was more than thirty-three thousand. Without immediate action, drug-resistant infections will kill ten million globally every year by 2050. That's more than die of cancer. And more *will* die of cancer, because of this.[12]

The grim truth is that we're running out of drugs. The World Health Organization puts it bluntly: "The clinical pipeline of new antimicrobials is dry." Otto Cars, professor of infectious medicine at Uppsala University, tells me: "We don't even know if we can get new antibiotics." And the major driver of this victory of superbugs over human innovation? Misuse of the drugs we already have.[13]

I don't want to imagine a world where even minor wounds mean hospital stays, when chemotherapy is unviable and even a slight infection is a death sentence. But if we don't get our collective act together, that could be the one our children inherit.

PRIVILEGE AND RISK

Amid these fears, I recognize something else as well. I face this moment from a position of comparative privilege. It may come as a shock to some parents that governments are failing catastrophically to safeguard our kids' futures. But for many communities, this is nothing new.

I am reminded sharply of this when I read an open letter from writer Mary Annaïse Heglar to the climate movement: "I'm with you when you say that climate change is the most important issue facing humankind. I'll even go so far as to say it's the most important one ever. But when I hear folks say . . . that the environmental movement is the first in history to stare down an existential threat, I have to get off the train. . . . For 400 years and counting, the United States itself has been an existential threat to Black people." Heglar forces us to acknowledge a stark and uncomfortable truth. These global emergencies come on top of, and exacerbate, embedded patterns of discrimination and misrecognition which many parents, and children, have endured for centuries.[14]

I have a fragmentary, painful appreciation of this because my children are girls.

I tell them that their bodies are their own, that no one has the right to touch them without their willing consent. I tell them so knowing that 30 percent of women worldwide have been subjected to physical or sexual violence. That 27 percent of women aged fifteen to forty-nine who have been in a relationship have been assaulted *by their own partner*. That less than 3 percent of sexual assault charges result in a conviction in the United States and only 2.5 percent in a prison sentence. That sexual harassment is commonplace in UK schools. That the US Supreme Court has overturned *Roe v. Wade*, denying bodily autonomy to millions of Americans. That in 2016 a male judge sentenced a male Stanford University student to only six months in jail for sexually assaulting a woman student. The reason for this lenience? To minimize the damage to *his* prospects in life.[15]

I know that on March 3, 2021, Sarah Everard, a thirty-three-year-old Londoner walking home from a friend's house was raped and strangled by an off-duty police officer. That on July 29, 2021, Valerie Junius, a Texas mother of six, was shot, apparently by the husband she was planning to leave. That in 2018, 1,946 women or girls in the United States were murdered by men or boys.[16]

I feel all this. But there is much that I am fortunate enough *not* to feel, in the same immediate way. My daughters inherit the social disadvantage of being girls, from being routinely objectified to gender pay gaps, but they have stepping-stones to advantage, too: whiteness, middle classness, nondisabled bodies.

In 2020, Kenya Young, a Black woman and NPR executive producer, was telling her sons, "I don't know what to do anymore, you guys. I don't know how to save you. I don't know how to keep you safe. I don't know." She nursed her third baby crying because two more Black men, Philando Castile and Alton Sterling, had been killed by police. When her sons go to the park, she gives them the "talk." Not the one that white parents give: "Wear sunscreen, don't talk to strangers, don't climb too high." *This one*: "Don't wear a hood. Don't draw attention to yourself. If you get stopped, don't run. Don't sign a confession."[17]

Don't get noticed.

Racism kills. The *Washington Post* has logged fatal shootings by on-duty police since 2015 and the conclusions are damning: Black people are killed at more than twice the rate of white people. Racism makes Black kids sick, from birth disparities and maternal stress to mental health problems and chronic disease. Black *babies* are more likely to die than white ones. Racism also makes Black kids poor. Seven in ten attend high-poverty schools, with significantly worse academic outcomes.[18]

The future is clouded for all our children, but the present is already hard enough for those in marginalized communities: Black kids, children of color, Indigenous kids, trans children, gay children, or children with disabilities. (Another grim statistic: in the twelve months leading up to spring 2017, one in five LGBT people in Britain suffered a hate crime or incident because of their sexual orientation or gender identity.) What's more, these hazards are what philosophers call *intersectional*, cutting peculiarly into the lives and prospects of anyone in more than one disadvantaged group. We don't have the full story of Brock Turner, the Stanford University student let off so lightly for his crime, unless we also know that his victim, Chanel Miller, was a young woman of color, and that he and the male judge were white.[19]

My girls have another privilege, too: they were born into this beautiful city with its parks, good schools, fresh air, and effective sanitation, in an affluent country in the Global North. In 2011, 385 million children lived in extreme poverty, or less than $1.90 a day. In 2012, 168 million children were in child labor, and 85 million did hazardous work. In 2019, one in three children under age five suffered from malnutrition; 5.2 million died before reaching five years old, often from birth complications, pneumonia, congenital abnormalities, diarrhea, malaria, or infectious diseases. Often, all they needed was what the World Health Organization calls "simple, affordable interventions": vaccinations, food, clean water, trained health-care. Things I take for granted.[20]

In the face of disaster, these differences, across sex and race and many other categories, do not disappear under a convenient umbrella of "we're all in this together." Instead, they get bigger. Women and girls have suffered disproportionally from both climate change and lockdown, from mothers taking most of the career and mental health hit of homeschooling, to not having time for political participation, to domestic violence or being the first in families to go without food.[21]

While I worry from a distance about antibiotic resistance, many already live, and die, without access to existing drugs. The IPCC forecasts of droughts and floods, disease, food insecurity, malaria, and diarrhea are worst for Asia, Africa, and small island states, and especially for children. Climate change could increase the number of people living in poverty by 122 million. By August 13, 2022, 89 percent of people in Cuba were fully vaccinated against COVID-19, 75 percent in the UK, and 67 percent in the United States, but only 56 percent in Pakistan, 36 percent in Egypt, and 12 percent in Nigeria. Globally, the pandemic had forced an additional 150 million children into poverty by September 2020.[22]

In the United States, the COVID-19 death rate in the poorest counties is almost twice that of the richest ones. As of July 2022, a Black or African American was 1.7 times as likely, an American Indian or Alaskan Native 2.1 times as likely, and a Latino person 1.8 times as likely to die from COVID-19 as a non-Hispanic white American. In the UK, demand for food banks (a sure marker of desperately struggling families) has soared since the start of the pandemic.[23]

All this should (and does) sit uneasily with me. Still more uncomfortably, I know that many white families, like mine, have benefited from the centuries of history that created these divides: simultaneously undermining the atmosphere for generations to come, and systematically abusing many of our fellow human beings through slavery, colonialism, and their still-tangible aftermath. I know, too, that as things stand, my daughters and I will use a disproportionate share of what natural resources remain.

My discomfort isn't the point. The point is that if I want to be a decent human being, as well as a good parent, I must remember all of this. I must respond appropriately, even as I worry for my own girls, and strive to protect them.

Let's talk about what that means.

I   OUR MORAL MISTAKES

1 BEING PARENTS AND BEING HUMAN

Before I was a mother, I was a philosopher. As such, I can offer no cut-and-dried answers to every quandary, but something that should ultimately be more helpful: a moral crib sheet, so you, my reader, can figure things out for yourself.

Moral theories are lenses through which to view the core problems we face, as parents and human beings. Rather than adhere to one philosophical worldview, I will build up my crib sheet, using clear reasoning, from a handful of ideas we can treat as a cornerstone of "commonsense morality." In other words, from shared ground that many philosophers, across different perspectives, would accept—and that, frankly, most *people* would agree on. As we go on, I'll sometimes delve further into one perspective or another, to shed light on particular dilemmas.

Some philosophers think morality comes down to maximizing total or average well-being. We call them utilitarians. Some (act utilitarians) apply this to individual acts. "Will it make more people happier if I buy this dress, or write a check for Oxfam?" Others (rule utilitarians) think we should each obey the moral rules that would increase overall well-being the most if they were the social norm; for example, "Don't waste water, because if no one does, then we'll all get through this hot summer."

Others, following Immanuel Kant, think we can derive morality from rational consistency. They think we should only do something if it's in line with a moral principle we can, without logical contradiction, want everyone to live by. On this reasoning, I mustn't tell a lie, even if it would help me out, because if everyone lied for convenience, there would be no institution of truth telling, and *my* lie couldn't help me.[1]

A third school of thought starts neither with well-being, rights, or duties, nor by judging individual actions. It starts with character traits. This is called *virtue theory*, inspired by the ancient Greek philosopher Aristotle. From this perspective, what matters is the kind of person I am, not what rules I follow. If I am virtuous, I will be truthful, temperate, benevolent, and so on. Still others—an increasingly influential set of feminist thinkers known as care ethicists—begin somewhere else again. They think caring *relationships* are valuable in themselves, to be cultivated and protected. Parenting is, of course, a core example.[2]

I start this personal and philosophical journey with three simple but powerful moral ideas that make sense on most of these schools of thought. First, there is such a thing as a "decent" human life: the kind of life we want, as a minimum, for our kids and ourselves; the kind we can assume other people want, too. Second, there are some basic things that everyone should or should not do for, or to, anyone else. Philosophers call these universal moral duties, positive and negative. Third, there are specific things we should each do for certain people. These so-called special duties are owed to our partners, parents, friends, colleagues, or compatriots. Most of all, they are owed to our children, because they are our children.

A HUMAN LIFE

Suppose you could have your child plugged in for life to a virtual reality computer game. In this artificial world, they would believe themselves to be completely happy, having amazing experiences. In the real one, they would be in a small room, being fed through tubes. Would you say yes?

I wouldn't, any more than I would choose it for myself. More than that, I would consider it an incredible betrayal of my potential-filled girls to sign them up to this chimera of enjoyment: a future in which, as the philosopher Thomas Hurka puts it, they would have no knowledge of the world or their place in it, no genuine achievements or real relationships.[3]

I don't want my children merely to believe their *preferences* are satisfied, any more than I want that for myself. I don't even want them to have all their preferences satisfied, their well-being weighed according to some sum-total of perceived wants. I want them to be happy, but I want that happiness to be the lasting satisfaction of a life fully lived. Either there is more to happiness than subjective experience, or there is more to a decent human life than happiness.[4]

We need a working definition of "human *flourishing*," or what it means for our individual lives to go well. We need it to make sense of what we must do for our kids, and what we must do (and not do) for everyone else. But, in finding that, we must avoid two dangers: the one I just highlighted, of thinking only purely subjective welfare matters, *and*, at the other extreme, the danger of being too rigid about what is required. If a "decent life" is defined too narrowly, it leaves no scope for our children to be themselves or to live among others who think differently from them.

Fortunately, there's a compelling middle ground. It was developed by leading philosopher Martha Nussbaum and development economist Amartya Sen, and it's broadly in line with the human, and sustainable, development goals. It looks like this.

We all have basic needs. We need to be healthy and sheltered, fed and watered, free to move, spared pain. But that's only the baseline. A fully human life is a life we "have reason to value." That means, says Nussbaum, being able to reason, think, and express ourselves, to use and enjoy our senses and imagination. It means reading, writing, dancing, singing, or having "down time." It means being able to seek religious or spiritual fulfillment, in your own way. It means being able to plan your own life and play your part in the decisions that determine how that life will go. It means not being thwarted by crippling fear or anxiety. It means being able to love and be loved, care and be cared for, enjoy self-respect, show empathy and concern. It means being able to grieve and feel grateful.[5]

That is what I want for my children. It is what I want for myself. However, I am not only an individual with my own interests and relationships. I am also a moral agent, to whom universal moral rules apply. As such, I am obliged (on almost any moral philosophy you want to subscribe to) to think not only of my own flourishing, or even that of my daughters, but also of our impact on those around us. This is too easily, and too often, forgotten. But it's still true.

What does it mean? Well, we can start with the fundamental Hippocratic injunction: do no harm. This doesn't just apply to doctors; it articulates an intuition without which we could hardly be said to be moral beings at all. More specifically, *don't seriously harm another human being*, if you can avoid it. Don't kill them, maim them, make them ill, take away their children or their home.

This "no-harm principle" makes sense in duty-based (or what philosophers call "deontological") terms because it is a cornerstone of respecting our fellow humans. I must, for logical consistency, want everyone else to follow this rule. However, it was formalized in Western philosophy by John Stuart Mill, often regarded as a rule utilitarian, and it makes sense in those terms too. We are all better off, overall, if everyone sticks to it. Although virtue theory focuses on character traits rather than actions, a virtuous person will characteristically *behave* virtuously. If you're not cruel, you don't go around stabbing or starving your fellow humans.[6]

Commonsense morality tells us this, too: *if someone is in desperate need, help them, if you can do it comparatively easily*. This is a moderate version of what the philosopher Peter Singer calls the "principle of beneficence." It, too, makes sense on more than one moral perspective.[7]

If you're a rule utilitarian, you reason that people will be better off overall in a society whose more affluent members protect the most vulnerable. If, like Kantians, you recognize yourself as someone to whom moral duties apply, the suffering of your fellow humans *must* matter to you. "Would [the virtuous person] help the wounded stranger by the roadside . . . or walk by on the other side?" asks philosopher Rosalind Hursthouse. "The former, for that is charitable and the latter callous." Benevolence, too, is a virtue: if not one of Aristotle's, then at least widely recognized by later virtue theorists.[8]

As a basic moral rule, this is also deeply, intuitively compelling. Take Singer's own poignant example. You see a drowning child on your way to work. You could rescue them, but you'd ruin your new shoes. Should you do it? Show me the person who says no, and I'll show you a sociopath.

So far, so simple. But we have special ties to some of our fellow humans, and new obligations to match. Most of all, when we have kids, everything is a hundred times more complicated.

SPECIAL TIES AND SPECIAL DUTIES

When my oldest daughter was five days old, she had to go to the hospital. She was dehydrated. For twenty-four hours we waited, terrified, for her to pee. Nine years later, she is strong and boundlessly energetic, but I have never stopped feeling terrible about it. I still cannot look at photos of her from those first few days. One of the most fundamental things we owe our children is to give them nutrition. My failure, however unintentional, cut at my heart. It still does.

A few days after our return home, I went out alone with my baby for the first time. We were meeting the friends from my antenatal class: a group of amazing women who would, over months and years, become my therapists, friends, and constant support. I remember how I fumbled, wrestling with the unfamiliar folds of the sling. I remember how my own mother lifted my tiny daughter, supported her against me, until I learned to get it right. I remember leaving Mum, stepping with infinite care down the worn stone steps from my tenement flat. I remember the bus ride, walking along the street and into the café, holding my child, warm and perfect and vulnerable, against my heart.

I was terrified of dropping her. I still am.

It matters to us to do good to our children. In this reality lies much of the joy, but also much of the fear, of parenthood. Behind the fragile laughs and tearful confidences of those get-togethers with other mothers lay an all-consuming dread of getting it wrong. As a philosopher, I can put this more strongly. Over and above whatever we should do for all our fellow humans, we *owe* it to our kids to care for them and help them do well. Even when the emotion is missing or misdirected—and it can be—the parental duty is nonetheless real.[9]

Here's one explanation, going back to that least controversial of moral rules: don't harm others. We can become responsible for protecting people *because* we have hurt them or put them at risk of harm. If I knock the roof off your house, the least I can do is keep you sheltered from the rain. Most parents cause their children to exist and simply *by* existing, they are made incredibly vulnerable. As babies, they are most evidently so because they cannot do anything for themselves. But it goes well beyond that. Children have needs, which must be secured for them. Adults have needs, which they must learn as children how to meet. We bring our children into the world; we must not leave them out in the storm.[10]

However, not all parents are biological parents. Adoptive parents are parents; stepmothers and stepfathers are parents. Love does not hinge on biology or, for that matter, on procreation. Nor can parental responsibility depend on it, or very many parents would owe their children nothing at all. What's more, some biological or procreative mothers and fathers are parents only in a technical sense. Sperm or egg donors, surrogate mothers, or those who give up newborns for adoption fulfill their obligations to the vulnerable beings they have helped create by ensuring that others will take them on.[11]

So here's another explanation: we take on parental duties like we take on many other obligations: voluntarily, or intentionally.

Standing before the celebrant, I promised my husband to love and cherish him, for richer, for poorer. Signing my employment contract, I made a commitment (moral as well as legal) to educate my students. Over time, friends come to understand that they can rely on each other, they start reasonably to expect it of each other, and, through the same process, to have a special responsibility to each other. In becoming parents, we commit to caring for our children. Often, we do it explicitly. Some stepparents make wedding vows to their partners' children. My husband and I held naming ceremonies for our babies, promising each wriggling infant, in front of family and friends, that we would do our best for her. But we had already made that promise implicitly. We all make it when we take on the task of parenting, whether we sign adoption papers or hold our bloodied newborns in our arms, and don't let go.

We might go further. Harry Brighouse and Adam Swift, both fathers as well as leading philosophers, think this isn't just a commitment we choose to make, but one we must make if we want to enjoy the incredible experience of being a parent. We determine our children's destiny, to a greater or lesser extent, and we do this not as lawyers or doctors or teachers, but through a deeply absorbing shared life. This dangerous power comes with a moral condition. We must use it to serve *their* interests.[12]

But what does that mean?

According to a survey of teachers, support workers, pediatricians, and health visitors, "good enough parenting" means meeting our kids' health and developmental needs, putting those needs first, providing reliable, consistent care, acknowledging problems, and engaging constructively with support services if needed. That's a lot. But I want to do *more* than enough not to attract the negative attention of the state, since I'm lucky enough to be a position to do so. I want a moral job description.[13]

BEING GOOD PARENTS

"Pretend this . . . ,"my older girl says. "No," the little one says—interrupting, bouncing with enthusiasm—"'tend that . . ." I smile, listening, because decades ago my sister and I thrived in just such worlds of our own creating.

Childhood matters in itself: it matters in a way that can never be compensated for, no matter how good the adulthood. Our children are particularly vulnerable in childhood, needing warmth and shelter, food and care, and needing us to provide it. But there are also needs, like play, which are arguably peculiar *to* childhood. Through play, unstructured and all-absorbing, children can transform reality. They can shape it and symbolize it in their own way to meet their own psychological needs. And play is fun.[14]

But our children won't stay children.

My youngest is six as I write this. She teases me with a litany from which she derives apparently endless glee. "This year I'll be seven. Next year I'll be eight and the year after I'll be nine and . . ." And so on, all the way to eighteen, and beyond. I squeal in protest. I kiss her and tell her she'll always be *my* baby. But it's also my job to help her grow up, and I know it.

Good parenting is about equipping our kids to lead the kind of full human life I described at the start of the chapter. For this, they need to develop physically, cognitively, socially, emotionally, and morally. They need to learn to think rationally, make choices for themselves, and make them meaningfully and responsibly. They need help to recognize and work with their emotions, and to develop caring bonds. Another reason play is so important is that it helps kids grasp increasingly sophisticated concepts, manage their behavior, build social skills, and practice them.[15]

This is a lot. So let's be clear. I'm not saying this is *only* our job. Schools, teachers, governments, wider family, councils, religious leaders, community organizations, and others, should all help our children along this path. However, the moral buck stops with us, as parents. This is partly because of the commitment we made, or the fact that we brought these new people into this uncertain world. It is also because, in practice, it cannot be done without us.

According to Brighouse and Swift, children need someone to love and raise them, not just impersonal care. They need someone to respond to them, for emotional and social development. They need someone who communicates with them one to one and who teaches them, through that developing communication, about themselves and the world. Brighouse and Swift think this makes the close-knit family unit superior to other ways of raising kids. I don't know if they are right, but I know we *do* organize ourselves into these units in societies like mine. In practice, if our children are going to get the love and consistency, the reliable communication and responsiveness they need, they'll have to get it from us.[16]

We might reach the same point by another philosophical path, as care ethicists do. Rather than value this relationship instrumentally, because it helps our kids, or us, to flourish as individuals, we might instead recognize that we are essentially dependent beings and that caring relationships matter in themselves. Then we could treat the parent-child tie (particularly the *mother*-child tie) as a core example, perhaps the paradigm case of just such a relationship.[17]

Either way, the relationship matters. A lot.

BEING GLOBAL CITIZENS

We live in an interconnected and multi-threatened world. As a moral agent, like it or not, I am also a "global citizen." So are you, and so will our kids be. Let's talk about what that means.

Consider Jdimytai Damour, a thirty-four-year-old store worker for Wal-Mart in Long Island who wrote poetry and wanted to be a teacher. On November 28, 2008, two thousand shoppers forced their way into his store for the "Black Friday" sale, knocking him down and trampling him. He died because they wanted first dibs on digital cameras and plasma TVs.[18]

A horrible story, but I'm telling it for a reason. It's no longer enough to think only of the harm *I* do. I also have to consider what *we* do. In more philosophical terms, the no-harm principle can be *collectivized*.

In individual cases, you are responsible for the harm you cause if you knew, or should have known, that what you were going to do would seriously harm someone, and you could have acted otherwise. Damour's death was avoidable. Not one of those bargain hunters *had* to be there. They each wanted to act in a certain way, they each knew lots of others wanted to act that way too, and it was obvious that someone could get badly hurt if they all did. We can call this a shared harm; we might call it a kind of collective responsibility. Either way, those shoppers will have to live with what happened that day.[19]

I'm not here to judge them. I'm here to do something that will make most of us feel still more uncomfortable: reflect on how many of us are *like* them. Our actions combine in predictable ways to do terrible things. We talked about them in chapter 1: climate change, antibiotic resistance, pandemics, searing poverty, and institutionalized discrimination. These are what the philosopher Judith Lichtenberg calls "new harms": complex, global harms. They are real, and they are everywhere. How many of us would take candy from a baby, when your hand must take it from theirs, let alone the only meal they are likely to get? But that's what we do between us, all the time.[20]

The climate ethicist Stephen Gardiner puts this another way. When we drive rather than get the bus, eschew Rainforest Alliance coffee, buy supermarket baby clothes, cook that Sunday joint of beef, book flights for a family holiday, we are, he says, playing a part in a global pattern of trade, consumption, and greenhouse gas emissions. In other words, we are upholding social and economic institutions that *we know* destroy other people's lives. Even extreme weather, once categorizable as an "act of God," can be traced back to affluent lifestyles in affluent countries.[21]

This is not about *blaming* you (or myself) for something we did not individually control. I'm not saying we (or those Black Friday shoppers) should feel guilty. But shame, says the philosopher Larry May, is a "subtler pressure," implying association rather than individual causation. You can be ashamed of what you are a part of.[22]

Perhaps you find this unconvincing (as well as uncomfortable). I hope not. But if you do, try thinking about it like this instead. Forget whether you and I share responsibility for these terrible harms. Just admit that they *are* happening and think of the good we could do together.

Picture this. August 20, 2020. Durdle Door, on the southwest coast of England. A jagged arc of limestone, sprayed by wild sea. A beach with a throng of sightseers. A beautiful place with a fatal secret: a vicious riptide. A swimmer, caught out, struggling in the water. Another man, swimming out to rescue him, also buffeted by waves. And a solution: a human chain of beach goers, hand in hand, stretching into the rough water, hauling them to safety.[23]

That's a true story. Life imitating philosophy. Or, happily, *not* quite imitating it, for twenty-eight years earlier, May described a hypothetical scenario: a swimmer drowning; beachgoers, uncoordinated, failing to organize themselves in time. Those beachgoers, he said, shared responsibility for failing to do what they could easily have done together, to save a life.[24]

The point those Dorset beachgoers instinctively illustrated is this. If you and I and others can spare someone from terrible suffering by cooperating, and we *can* easily and safely cooperate—take hold of each other's hands, step into the water, form that human chain—we should do it. Accept the basic moral idea that each of us should help the desperate, then it is inconsistent not to accept this too. Philosophers debate where exactly cases like this lie, on the line between individual and collective wrongdoing. For us, it is enough to know this. If those beachgoers had not acted as they did, the families of those who died could have said, with rightful anger: "*You* could have saved him." In English, we don't distinguish our "you" singular and plural. But it would be, in this case, firmly plural. And it would make sense.

Now look forward. We live, literally and figuratively, in a world full of beaches with people drowning off them. Of course, there are many others—the richest of the rich, governments, corporations—who have a more obvious responsibility to redress these harms. But they *don't* redress them. While I was writing this, in the November cold, twenty-seven migrants drowned in a deflated boat in the English Channel. Politicians rushed to blame each other. Suppose there had been a lifeguard at Durdle Door, who threw up his hands and walked away. He would have acted wrongly, as well as unfairly, but the others should still have done what they safely could without him.[25]

Resolving global crises like climate change requires (obviously!) much more than small-scale collective action. But that doesn't let us off the hook. It means, says philosopher Henry Shue, that we should organize ourselves to challenge, reform, and re-create our political, social, and corporate institutions.[26]

Citizenship is a sophisticated and contested concept. But for my purposes, the point is simple enough. Being a "good citizen" is about more than obeying the law, playing your part in maintaining just social and political institutions. It also means doing your bit to make sure that those institutions *are* just: that, at a bare minimum, they protect rather than destroy the lives of those who depend on them. Whatever else being a "good global citizen" means, it means responding appropriately to the *global* crises that we are a part of causing, and could help end.[27]

The problem, of course, is that this takes a great deal of time, money, and emotional energy. How do we do it, and still do everything we should, for our own kids?

Chapter 1 in Essence

As human beings, we are moral agents. This means we have some responsibilities to other people, wherever and whoever they are. We must not seriously harm them, and we should save them from great suffering if we can do it relatively easily. That's commonsense morality; it also fits most philosophical perspectives.

As well as these universal moral duties, we have special obligations to certain others, especially our own kids. We make a commitment to them, implicit or explicit, when we take on the amazing but asymmetric relationship of parenting. Often, we bring them into the world, vulnerable to its escalating hazards. That means, at a minimum, that we should meet their needs as children and bring them up able to lead a "decent" human life as adults. That's not necessarily a life with lots of money, nor even one where they feel content, day by day. It's a life where they can enjoy central human interests. They need to be able to be healthy, plan their own lives, enjoy their senses and imaginations, love and care for others.

In our interconnected, imperfect world, our moral duties collectivize. We should work together to prevent the harms that we exacerbate between us: harms like climate change, antibiotic resistance, pandemics, institutionalized discrimination. We should build institutions to protect the vulnerable. This, at a minimum, is what it means to be good global citizens.

2 OUR CHILDREN AND DISTANT STRANGERS

On September 2, 2015, a three-year-old Syrian boy, Alan Kurdi, drowned in the Mediterranean Sea. Desperate for a livable future, his family had paid thousands of dollars to board an inadequate inflatable boat to the Greek island of Kos. They never got there. On October 26, 2018, seven-year-old Amal Hussain died of starvation in a refugee camp in Yemen. She died not at the arbitrary whim of fate, but because of food shortages perpetuated by a war on Yemeni civilians, using weapons sold by the United States, France, and the UK.[1]

The stories are not unusual: they are, horribly, the reverse. But those two names are familiar to me, unlike those of countless others, because their plight caught the attention of the world media. The children's photos appeared on newspapers, televisions, and computer screens across the Global North. A tiny body in a red T-shirt and dark shorts, face down in the shallows. A child all bone and sinew, days from death, her eyes looking away from the camera. The privileged world exclaimed at these images of tragedy. We wept, hugged our own children close, made some hasty donations. Then we moved on, got on with our days, and went back to piling comforts on ourselves and on our kids.

In the United States, the average child will get $6,500 worth of toys over their lifetime. In 2021, three-quarters of parents polled by BabyCenter.com had spent more than $100 on their baby's *first* birthday: one that (presumably) the child will not even remember. Super-wealthy parents spend tens or even hundreds of thousands on elaborate "soirees" for their toddlers. Linden Hall School in Pennsylvania charges $60,000 a year in fees; the UK's Eton College nearly £45,000.[2]

Day to day, many of us treat our children to meals out or overseas holidays. We give in to the incessant demand for yet another plush toy, sequined top, or Xbox game. We do it knowing that other children are dying for want of basic necessities and, often, we don't feel bad about it. In 2020, a British mother was splashed all over social media after posting the huge pile of Christmas presents she had gotten for her child. She was unlucky, in the public attention she attracted, but she's not unusual. "We spend hundreds," one commentator said, defending her, "and there's not a single person who would be able to make me feel bad for that." "So what," said another, "it's her kid, she can buy what she wants for HER CHILD!" But the truth is, it's not that simple.[3]

When we have kids, it's tempting to think that gives us a kind of moral "free pass" from fulfilling many of our responsibilities as global citizens. We think it's OK to spend all we have on toys and books and gadgets, birthday parties and holidays, sports coaching and job or university applications. More than that, we sometimes think it's what we *should* do. It's not selfish or immoral; it's part and parcel of being a "good parent."[4]

In thinking like this, we make two serious moral mistakes. We overestimate some of the things we should do for our own kids, when others need us too. And in the process, we let down the very children we are trying to protect, because we're mistaken about what *that* means, in this fragile world.

THE DEMANDS OF MORALITY

"I don't know whether there are any moral saints," writes philosopher Susan Wolf. "But if there are, I am glad that neither I nor those about whom I care most are among them."[5]

There is a limit to what's expected of us, as moral agents. Many philosophers (and, again, most people) would agree on that. I can simultaneously acknowledge that human suffering is terrible, and I must do something about it, and believe that I am entitled to *some* space to lead my own life and nourish treasured relationships. Space to rest, to ride my bike, to write fiction and listen to music. To laugh until I cry with my sister or climb mountains with my husband. To applaud my daughter's handstands, walk along beaches with my parents, or play board games with friends.[6]

Of course, we must not do serious, individual harm, however hard we find it not to. We must not kill someone or make them sick. In philosopher-speak, this is an extremely demanding duty. But when it comes to our duties to help others, most philosophers let us off the hook more easily. A life entirely dominated by altruism has no space for other valuable things; a society where all attention is devoted to helping the needy is a society without the music of Mozart or Leonard Cohen, Shakespeare's plays, or Toni Morrison's prose.[7]

In a better-organized society, within a just global order, we would be spared much of this individual soul-searching, because the ongoing crises that mark our perilous world are not, usually, the direct result of individual actions. They are harms you and I are part of, but cannot prevent simply by *not* being part of them. Just institutions would prevent them on our behalf. They would protect the vulnerable, future generations, the world we all depend on. We could be "good global citizens" simply by following the laws and norms of our societies. Then we would be free to focus on our own immediate projects, with whatever space and money we had left. Our valued pursuits. Our interests. The people we love.

Unfortunately, that's not the situation we're in now. So how do I, or you, strike this balance? How much space is enough?

At a minimum, we should be part of this action to save untold lives, and to stop destroying them between us, if the cost to us is minimal. But given what it is at stake, that doesn't appear to be demanding enough. Even if we can stop short of significant sacrifices, surely more is required than the merely trivial ones.

This seems clear, but things are more complicated in practice. Even trivial sacrifices can become significant if you are making them continually. Take the intuition that we should save someone from great harm, on any given occasion. As Singer and another philosopher, Peter Unger, remind us, we can do this at *any time* with a bank transfer to a reputable aid organization. Do I have coffee with a colleague next week, or give a family safe water to drink? Does it matter more to read this one story to my daughter than send emails recruiting more climate activists? Iterate this, and my life is full of such choices.

I could end up never being able to do the small things that, in combination, make up my life and relationships.[8]

It helps to think of the sacrifice we should make for morality in a different way: in terms of its impact on our lives as a whole. Perhaps I can "ring-fence" certain commitments—valuable pursuits, my own core interests, important relationships—but also think long-term about *how* I pursue them. There's a big difference between giving up some key part of my life, even temporarily (what we might call a "significant" cost), and finding a less expensive, wasteful, or time-consuming way of enjoying it. If I can strike this balance, I can set aside the time and effort and money needed to maintain my friendships, family ties, and life projects adequately overall. I can be a good global citizen without weighing every individual choice: buying an ice cream cone for my kids, or sending more money to UNICEF.[9]

One of our most central commitments is the one we make to our own children. But perhaps "adequately" is the key word here, too.

DEVOTION WITHOUT DROPPING EVERYTHING

Between 2014 and 2016, researchers for the National Academies of Sciences, Engineering, and Medicine reviewed published material on child development. They found that parents should follow their kids' lead, respond predictably, be warm and sensitive, have rules and routines, share books, talk with them, keep them healthy and safe, and use appropriate (not harsh) discipline. Around the same time, psychologists Robert Epstein and Shannon Fox studied data from two thousand parents who had taken an online test of parenting skills. "Good outcomes" were a high-quality parent-child relationship, happy children, and healthy children. The best predictor of these was "love and affection." Nothing, there, about custom-made, $5,000-dollar ball pits.[10]

But let's dig deeper.

We owe it to our kids to care for them in childhood and set them up for a decent adult life. We saw that in chapter 1. We can't legitimately *be* parents without doing this. Even beyond what we owe our children, we generally want to do more for them than we do for others and are often prepared to give up a lot to do it. This partiality is understandable. It is grounded in a relationship that is itself valuable. Does that make it OK to be "parental saints" (or, depending on how you see it, "parental villains"), sacrificing everything else to make our children's lives better?

No.

Point one. Our other interests, projects, and relationships don't vanish into thin air, just because we have kids. They still matter to us, and *we* still matter, too. If thinking that way feels selfish (it shouldn't, but it might), then put it like this instead. Childhood matters, but adulthood matters too. Our children, like us, have whole lives to lead. Suppose we ask, with philosopher Matthew Clayton, which of these scenarios is better *overall* for our children: One where social norms allow us some scope to lead our own lives now, and do the same for *them* if they are in our shoes? Or one where we must devote ourselves 100 percent to our children and they must do the same for any children they have? A society of "parental saints" doesn't sound much more appealing than one of moral saints.[11]

Point two. We don't stop being moral agents, either.

In Guillermo Martinez's novel *The Oxford Murders*, a man crashes a bus full of children with Downs syndrome to get a lung transplant for his dying daughter. We can be unequivocal about his action. It's wrong. Gut-wrenchingly so. He owes a lot to his child, he loves his child, but neither of these truths can override the fundamental duty not to kill others.[12]

When it comes to our role as global citizens, things are less clear-cut, but our broader obligations still matter. Perhaps our duties to our children can override the claims on us of those in desperate need, or at least some of them. We take on commitments to our kids, as part of the complex and incredible relationship we have with them. They depend peculiarly *on us*, as individuals, and we are peculiarly responsible, in many cases, for the situation they are in. But we owe them the opportunity for a "decent life," not the best possible one. There's no reason to assume that our desire to do more and more for them, even after that, has the same moral clout. On the contrary, drawing the line between global citizenship and this kind of parental partiality looks much like drawing the line between a major moral duty and any other interest or commitment we might happen to have.

Think like this, and most of us are drawing that line in the wrong place.

MONEY WON'T BUY YOU LOVE

My father read to us as children, perched on the end of the bed with a book by C. S. Lewis or Arthur Ransome. I remember it vividly: my sister, close beside me; the way he would stop after a chapter or two and pretend my toy tortoise was attacking us, until we were helpless with giggles. I still have that tortoise, and I love reading to my daughters. I love them snuggling in, hiding at the scary bits, jumping with joy when it's all OK.

Here's one reason for partiality. The parent-child bond *matters*. Whether we start from what children need, *or* from caring relationships as valuable in themselves, this one is part and parcel of being a good parent. "If people were plants," says psychotherapist Phillipa Perry, "the relationship would be the soil." Some activities are central to maintaining this relationship, so we *should* make time to do them with and for our children.[13]

But the archetypal "relationship goods," say Brighouse and Swift, are reading bedtime stories and doing things you value together, as a family. Neither of these require endless resources. It's *quality time* that matters, not exclusively hired theme parks or five-star holidays. For me, sharing something valuable with my kids means walking, climbing, camping, or even going on climate marches together. For others, it's attending church or mosque. For others again, maybe it's watching football or visiting art galleries.[14]

The relationship is key, but it isn't the only thing our kids need, or the only thing we think it's important to give them. Another reason we pile resources on our own kids is that we think it's good for them in other ways.

At one level, this is straightforwardly wrong. It's *obviously* bad for kids to be gratified in every material whim. That's not sophisticated morality; it's basic parenting sense. I'll admit, I've taken the "easy" option plenty of times: "OK, you can have an ice cream/buy that hideous plastic toy/go to the playpark. Just please, please stop screaming." But if every time my child wailed "It's Not Fair" I rushed in to make it (what she thinks is) fair, I'd be doing her a disservice, as well as everyone else. It's far more important to teach her to deal with setbacks for herself. It's developmentally important, say Brighouse and Swift, for our kids to see and relate to us as real people, with interests, aims, projects, and other relationships that go beyond them. That means us being able to keep on pursuing those other goals, at least "to some extent."[15]

For further insights, I call Cathrine Jansson-Boyd, a consumer psychologist at Anglia Ruskin University. She's informative, incredibly helpful, and, on one thing, unequivocal. "Material possessions," she tells me, "can really ruin children." Money may or may not increase reported happiness. The evidence is ambiguous. But depending on money for happiness is, undoubtedly, a problem. "The more materialistic children are in their teens," Jansson-Boyd says, "the more likely they are not to get a grounded idea of who they are." Whatever we want for our kids, it doesn't involve becoming increasingly insecure in adolescence or developing personality disorders.[16]

At another level, however, there's a lot that money, as well as time and effort, *can* do for our kids. It can give them experiences or help them develop skills that may, very plausibly, enhance their lives, such as travel, cultural experiences, sports and music lessons. To complicate matters further, money *can* buy advantage, and some goods are relative.

In October 2019, I dropped my then-five-year-old at school for Halloween. She was wearing a ripped bedsheet with not-very-neat holes for eyes: a costume I had created (a generous use of the word) in thirty seconds before leaving the house. We were surrounded by children in amazing outfits, apparently specially purchased or meticulously handmade.

My daughter didn't care. She swooped across the playground, shouting "Wooooo." But one day she will. And *I* felt like a cheapskate, a bad mum. I felt it despite everything I know as a philosopher. I felt it despite all that had been in my mind and heart, only weeks before, when I was researching for this book and came across Amal Hussain's photo, and her desperate story, for the second time.

Am I a hypocrite? I don't know. I know there is *some* method, in my apparent inconsistency, but it only takes me so far.

It is understandable to care deeply about the relative position of our own children—the *real to them* social woes—even while others suffer far more. Sometimes it's even appropriate to do so, since this, too, matters for flourishing. "The need for self-respect . . ." says philosopher Judith Lichtenberg, "is basic and universal. But what it takes to satisfy that need varies from time to time and place to place." Infrastructure makes it hard to live without something (a car) that you wouldn't need if things were better organized (decent public transport). Social life, work life, school life all rely on the latest technology. If our neighbors all wear shirts and ties, says Lichtenberg, we're ashamed to go out without them; if everyone else has a mobile phone, the same (at least arguably) is true.[17]

What's more, this is about much more than Halloween costumes. When the sons of billionaires buy their way into US politics or the UK cabinet is filled with the graduates of one university (Oxford) and just one private school (Eton, again), comparative advantage becomes a matter of paying school fees or buying a house in a privileged zip code, booking language or math coaching, or even using contacts from your own school days to get your kid an internship. And there's a natural reaction to all this. Given that the system is unfair, don't we owe it to our kids to make sure they don't lose out?

The answer is yes and no.

Yes, what our kids need to thrive will depend on where they live. But here's a thought experiment: a tale of two children. Anne lives in an entirely consumerist city. All pleasures must be paid for. All her friends have big houses and expensive holidays. Beatrice lives in a much simpler community. Time outdoors is relished. Community activities are free or cheap, celebratory meals prepared together. Does Anne *need* more than Beatrice? Arguably, she does. Her prospects in life, her social status, perhaps even her self-esteem depend on her being able to do at least some of the things that those her age around her do.

However, there are still things they *both* need, for any life at all. Even if it matters to Anne to have a party in the Trampoline Park, in a way that it doesn't to Beatrice, Anne doesn't need food more than Beatrice does, or a roof to sleep under. If Beatrice doesn't have those basic things, that should matter to *Anne*'s parents, as moral agents. It should matter to them even though they *also* (and appropriately) care about giving Anne the advantages she needs to get by in the society she will have to live in.

There's also an important distinction between a "fair chance" and a "competitive advantage," even if it's not always clear where one ends and the other begins. Our kids need a functioning life, from playground to workplace, but do they need to be better off than their peers? We don't like to think of ourselves as trying to make other people, other *children*, worse off when we pile advantages on our kids. As Brighouse and Swift point out, that would be "creepy." But in practice, says Swift, if we enable our own children to jump the queue, then we're pushing other children further back.[18]

Or we could think about this another, equally uncomfortable way.

SPENDING WHAT WE SHOULDN'T HAVE

Imagine this. Your neighbor buys a $5,000 playhouse for her kids. It's splashed all over Instagram with lots of "Nothing but the best for my little princesses!" How do you feel? Pleased for her children? Judgmental? Jealous? Worried your kids will want one too? Or like you couldn't care less?

Now suppose your neighbor stole that $5,000. Or that it was a legacy from her grandmother, who stole it. How do you feel now? I'm guessing "judgmental" is in the ascendant. Or imagine you know not only that she stole the money, but also that she took it from already poor families, whose kids now won't be getting any birthday presents.

Has that changed how you feel?

Here's another often-used justification for prioritizing our own kids over everyone else: "I can afford it." But maybe we should drill down on that a bit more, and think about *why* some parents can afford luxuries for their children, and others can't.

Consider this (because it's true). Some parents can afford a vast amount more than others because we live in a world that isn't even basically just. In a just world, we would have institutions in place to fulfill our basic moral responsibilities to our fellow human beings, and everyone would have the chance of a decent life. Instead, as we've seen, deprivation is widespread: not only in "relative" goods, but also in those most objectively necessary to live at all. This is true globally, and within states. In 2019, the richest twenty-two men in the world held more wealth than all the women in Africa. In the United States, in the same year, 14 percent of children lived in poverty. In the modern world, we tend to think of property rights as sacrosanct. But does anyone, parent or otherwise, have a moral claim to their money, *however* they got it, if they can build $5,000 playhouses while others starve?[19]

The philosopher Colin Macleod goes further again. He thinks those of us living comfortably in rich states are morally on par with the neighbor in my example. Of course, we haven't stolen in the legal sense to buy treats for our kids. But look closely at the history of this money we declare is ours and it gets very murky, very quickly. If mothers and fathers in famine-struck The Gambia can't feed their children, and many British families have plenty to spare, these aren't two unrelated circumstances. The UK is rich on the back of colonization that stripped African states—and people—of the wealth that could have been theirs.[20]

Or look closer to home.

"I borrowed five figures for college and nearly six for law school, including a high-interest private loan that my grandmother had to co-sign." That's Heather McGhee, former president of the Demos think tank and expert on racial economic and social injustice. "At forty years old, I'm still paying it off, and I don't know a single Black peer who's not in the same boat, even those whose parents were doctors and lawyers." Eight out of ten Black graduates had to borrow, she says, and they pay higher interest rates than any other group.[21]

There's a reason for that, and it isn't that white families worked harder for their kids' education than Black families did. The truth is as simple as it's grim. "Generations of racist policies have left our families with less wealth to draw on," says McGhee. Put it like that, and "It's *my* money" isn't the end of the conversation. It's only the start.

OUR CHILDREN'S FUTURE

Now for our second moral mistake: In doing what we think is right for our own kids—in prioritizing their immediate well-being, their material goods and economic prospects—we're also letting *them* down, because we're not protecting their future.

My children dream. They dream of going to France (twice postponed by the pandemic). They dream of skateboarding, surfing, and Harry Potter World. They dream of growing up to be an astronomer and an artist. Or perhaps an athlete and a vet. I read them books about space and animals. I encourage them to draw, to run, and to think. I find joy in their strength. I doubt myself when they struggle. I want them to be able to shape their lives for themselves. But they are growing up in a world that will constrain those lives, in many and terrible ways. If I want to be a good mother, I can't ignore that.

To flourish as humans, we don't just need to be able to do things, we need an environment in which we can do them. It's no use being a brilliant ice climber, without any snowy peaks to climb *on*. We need central opportunities, and we need them *reliably*. Our children cannot thrive, as children or adults, if they are constantly at significant risk of losing health or livelihood; if they eat today but might starve tomorrow; if fire or flood or looters could attack at any time. Even if the worst doesn't happen—and it *could*—the psychological damage is enough.[22]

Take this seriously enough—and I think we should take it very seriously indeed—and the conflict this chapter has taken for granted, between being good parents and decent human beings, isn't just overstated. It's partly illusory.

Here's what we might think—what those of us who have been cushioned by privilege probably *do* think: as parents, it's enough to focus on the kids themselves. We feed them. We give them space to play and time and toys to feed their imagination. We take them to school, or to the doctor's when they're ill. We read stories and nag them to do their homework. We referee sibling battles. We go to tennis matches and college shows. We listen when they are upset. Most importantly, we teach them to do things for themselves.

As for the rest—for an actual and political environment in which they can *have* a safe future, a society where they won't be murdered on the street or vilified for the color of their skin, air they can breathe without ending up in an ICU—we assume two things. We assume it's the job of our schools and local authorities, state and federal governments, and global institutions like the World Health Organization to take care of all that. And we assume that they will.

Unfortunately, that's a mistake. And that collective failure changes the rules of the parenting game.

"As a doctor for children what I do day-to-day at work is largely futile if these children have no safe future on this planet," says pediatrician Jenny Gow, one of the founders of the parents' climate action group Mothers Rise Up.[23]

If I owe my children anything at all, beyond merely keeping them alive *as* children, it is to protect them from climate change, antibiotic resistance, pandemics, discrimination, and the threat of war. *If* I can do so. That's not just me pulling rhetorically plausible responsibilities out of the air. It follows, clearly, from the uncontroversial obligations we already have, as parents.[24]

When I took my babies to be vaccinated, wincing at their cries, I didn't do it just to spare them *as* children from horrendous disease. I did it to protect them for the rest of their lives. Suppose your beloved child is at high risk of a debilitating genetic disease when she's fifty years old. Get her a vaccine now, and you protect her then. Don't, and she'll be powerless to undo that decision later. If you can do it, but don't, you're making a mockery of all the other things you do to prepare her for her future—the school runs and the vitamins and the fights over the iPad—because you're choosing not to make that future possible. You're reading bedtime stories in a house that's burning down.

That's what we're all doing, now. Except it's even worse, because not only are we failing to put the fire out, most of us are fanning the flames.

Moyagh Murdock, chief executive of the Irish Road Safety Authority, says congestion at the school gates is "incredibly dangerous" for small children. They weave in and out of the cars, easily unseen, easily hit. In 2018, UNICEF published a report, *The Toxic School Run*, which should have made middle-class parents squirm. Children in the UK are disproportionately exposed to pollution in the 7 percent of their day that they spend getting to and from school.[25]

Parents want to drive their kids to school—it's quick and convenient. They know other parents think that way too. They also they know that the traffic, with its attendant fumes, is killing children. They know it could kill their own kids. Again, our ambiguous English language lets us down. When a six-year-old is hospitalized for asthma, made worse by the exhaust of those school-run cars, no one can say to any one SUV-driving parent: "You [singular] did this." But they can say, "You [plural] did this," to them all together.

The same applies to comfortably off parents who fly on holiday, buy palm oil from deforested plantations, feed their kids antibiotics they don't need, or feed them animals who were given those antibiotics. It also applies to parents whose pensions or savings finance the oil and gas industries, and perhaps most of all, to parents who vote for the politicians who approve oil fields and subsidize factory farming.

PROTECTING OUR KIDS, TOGETHER

"We want a safe environment for our children," says Harriet Shugarman, activist and educator. "We baby-proof our apartments, but how do you do that when you step outside the door? How can you baby-proof the world?"[26]

A month after I first speak to her, I interview Lisa Howard, a PhD researcher at my own university who is studying the sociology of parent activism. In normal times, we're based in the same building, in Edinburgh's city center, but these are pandemic times and she's chatting to me from my computer. She tells me she interviewed twenty UK parent activists, trying to figure out their motivation. Her verdict? "Parents really wanted to be accountable to their children in the future. They wanted to look them in the eye and say they'd done all they could to address it."[27]

This makes philosophical sense. When it comes to protecting our children, the buck stops with us. If there's a giant hole in the pavement and the council has neither filled it in nor fenced it off properly, I don't say, "That's not my job," and let my daughter hurtle into it, at full pelt on her scooter. I keep her safe if I can. Then, if I remember, I email the council, pointing out that it failed to do *its* duty: the one it performs on all our behalf.

If our core moral responsibilities collectivize, so must our parental ones. If you owe it to your kid to give them a safe future, I owe the same to mine, and we can only achieve that by combining forces to give them both that future, what follows? That we should do just that. *I* can't baby-proof the world. Nor can Shugarman, or any of the activists Howard spoke to. Not while corporate giants continue to push oil and gas, industrialized agriculture, or mass consumerism; nor while governments, in supporting them, destroy the lives of future generations, distant communities, and their own marginalized citizens. However, changed institutions could provide that better future. Changed ways of living and raising our children together could; politicians who legislated for corporate change could. And we could do a lot, between us, to get institutions, social norms, and politicians to change.

I'll talk more later about what that means, for each of us. For now, it's enough to see this. In our multi-threatened world, being a good parent has a lot in common with being a good global citizen. On further reflection, they get closer still.

Chapter 2 in Essence

When we have kids, it's tempting to think we have moral permission to ignore the needs of strangers: we're not being "bad" global citizens, spending everything we have on our nearest and dearest; we're being "good parents." This is wrong.

First, we don't stop being human beings, just because we have kids. Yes, morality should leave us some scope for our own lives and relationships. As philosophers put it, our moral duties are not infinitely demanding. Yes, our duties to our children sometimes take priority over building a better world for everyone else. But these duties only take us so far.

Second, the way we live now feeds global emergencies, and these emergencies threaten our own children as well as other people. In other words, we're preparing our kids for adulthood, while undermining the world they will live that adulthood in. I can't address this alone, nor can you. But our duties to our kids, like those to other people, collectivize. We owe it, each to our own children, to work together to protect all their futures.

3 OUR KIDS ARE NOT ISLANDS

Here's one possible future. It's 2070. The earth is devastated, humanity is divided. The elite live in bubbles: underground cities, topped by domes. Your child, in his fifties now, is one of them. The cities are beautiful, created by the best designers, lit to mimic the changing colors of the sky. They won't last forever—maybe another half century—but they'll outlast your son. For the rest of his life, he will have artificially fresh air, fulfilling work, delicious food, incredible healthcare, art and architecture, beautiful libraries, state-of-the-art gyms. He can hear music in beautiful concert halls, watch sport in incredible amphitheaters. He can ski on pretend snow on fake mountains, swim in man-made versions of Iceland's Blue Lagoon.

The unlucky ones live and die at the mercy of the now-merciless elements, and there are far more of these people. They scrape a living from ground either parched or flooded, regularly attacked by cyclones. With neither vaccines nor antibiotics, they are decimated by disease. The animals live outside the bubbles, too, those that are left (for most species are extinct).

Almost all are livestock: farmed in huge sheds, the smell and heat and pain hidden from those protected elites who eat steaks and drink milkshakes. The last surviving predators attack vulnerable clusters of humans with increasing, desperate violence. The people attack each other, too, for they have been cast out into Thomas Hobbes's state of nature, and life is even more "poor, nasty, brutish, and short" than he predicted. Your child never looks beyond his sterile, dome-topped paradise. But he knows these desperate human beings are out there.[1]

Is this the future you want for him? It's not what I want for my daughters.

"No man is an island," the poet John Donne reminds us. And our kids are no more self-isolated beings than we are. They have their own interests, ambitions, and dreams, but they flourish also in relation to other people, places, even species. How well their lives go depends on more than their own well-being.[2]

In this chapter, I'll spell out what that means. I'll explain why truly being "good parents" means we must also be good global citizens, as well as good ancestors—and good ecological citizens. I'll explain what *that* means, too.

SHARED HUMANITY

We all share a threatened world, but we are not all threatened equally. *Children* are not threatened equally. If the introduction made anything clear, it was that.

Relatively affluent parents could take this in either of two ways. We could be like Xoli Fuyani, environmental education coordinator at the Earth Child Project in South Africa and foster mother to two eleven-year-olds. "I advocate for climate *justice*," she tells me, "because I know the social issues are just as important." Or we could be like Bob. I made him up, but he is, unfortunately, realistic enough.

Bob is a white man in the United States. He has a well-paying, secure job and a good pension. His kids, like mine, are white. Unlike mine, his children were born and identify as boys. They are straight, with no disabilities (at least that he knows about). Like me, Bob wants to be a good parent; he just has a very different idea of what that means.

Bob is worried about the prospect of global war, but only insofar as it threatens anyone on US soil. More precisely, insofar as it threatens people who live and think, and vote *like him*. He agrees that we need to prevent future pandemics in the Global North and tackle antibiotic resistance, but he's opposed to funding universal basic healthcare. Bob loves his sons, but he doesn't worry that they live in a sexist and racist world. After all, *they'll* be OK. The dice are loaded, but in their favor.

Bob accepts that we need to act on climate change. Perhaps he would even describe himself as "anxious" about it. If so, it's a very insular kind of anxiety: the kind environmental studies professor Sarah Jacquette Ray has in mind when she warns, "Climate anxiety can operate like white fragility, sucking up all the oxygen in the room and devoting resources toward appeasing the dominant group." Bob just wants mitigation, by whatever means will cost least to people like him, and adaptation for his own rich community. Beyond that, he thinks, it's not his problem.[3]

I think Bob is wrong. If we care about our children as well-rounded human beings, we owe them a world at least basically just: a world where *everyone* has a shot at a reasonable life, a world without extreme suffering. I don't want my girls to live in a sexist society because they're girls. I don't want them to live in a racist one either, even though they are white.

Here's why.

Over breakfast one morning, my eight-year-old asked me about airbags. What happened if they broke? Would you go through the windscreen? I told her I didn't know.

"Because it's never happened to you?" she said.

"No, it hasn't," I said, "and I wouldn't want it to. I especially wouldn't want to it happen to *you*."

My six-year-old stopped eating and burst into tears. "I don't want it to happen to *anyone*," she sobbed.

As emotional beings, the suffering of others matters to us. Not always, not consistently, but sometimes. Studies found secondary traumatic stress symptoms in high school children who had visited the Auschwitz Museum, and post-traumatic stress disorder in elementary school kids exposed to TV, internet, or newspaper images of death or injury in the September 11 terrorist attacks. The *New York Times* reported an outpouring of grief, after they printed Amal Hussain's picture.[4]

As moral beings, this suffering *should* matter to us.

We stand in relation to others, wherever and whoever they are. That's part of what it means to be human. Donne recognized it: "Any man's death diminishes me/because I am involved in mankind." I feel the truth of these words, written four centuries ago, without sharing his religious convictions. My little girl saw it instinctively, in her quick tears. We have moral reason to respond to others' pain—and not to be part of causing it. We saw that in chapter 1. Our children will be moral agents and global citizens, like us; they will have that reason too.[5]

What's more, the world we are creating for them is one in which it is very difficult to be both a morally sensitive being and a person with their own life and interests to pursue. White supremacy, says anti-racist author Jennifer Harvey, "malforms" our humanity. "None of us, regardless of our racial identity, can be truly racially healthy as long as we live in a racist society." I think this can be generalized. Just as we cannot flourish as humans without the background conditions to lead our own lives—a safe environment, security, access to food—so too, without the background conditions of justice, we cannot live fully in peace as fully human beings. Nor will our kids be able to do so.[6]

MORAL MARRING

"Sleeping difficulties, eating difficulties. Like, we spent days without eating because you just can't even swallow your own saliva. It's just, you can't even breathe normally." Those are the words of a healthcare worker who worked in humanitarian crises. They had seen their patients' needs clash with the needs of the organizations they worked for. They'd had to reconcile making ethical decisions with orders to the contrary from managers. They'd even had to choose which of two very sick children to treat.[7]

Consider that true story. Or imagine this one: a hypothetical example.

A woman lives on a wild bit of coast. Every day, ships are wrecked and passengers flounder in the sea. The woman spends her time either pulling them from the water or vainly petitioning for her government or neighbors to do so. She is exhausted. She barely sees her own family. Her friends tell her she has done her moral duty, has given up so much of her own life that she can legitimately stay home some nights instead. She knows this. And yet, if she does, she sees the bodies of victims, slamming against rocks, even as she holds her children in her arms. She knows she could have done more, even if no one (not even her) would say she *must*.

No one can have a duty to do the impossible. More than that, there are limits to what anyone can *reasonably* be expected to do. The healthcare workers shouldn't feel guilt for the choices they had to make. But they were in a horrible position. They had, say researchers Sarah Gotowiec and Elizabeth Cantor-Graae, experienced moral distress. That's the "painful psychological dissonance" that results when one's internal moral compass pulls one way, and outside limits pull another.

More philosophically put, we can be marred, as the woman in my story is, by choices between our own interests, our personal obligations, and the moral claims of others. According to some moral scholars, we can regret what we did, and regret it appropriately, *whichever* way we decide. Regret is a way of valuing the path not taken—more specifically, the path not taken *by you*. It does not depend on your having done wrong.[8]

Now think about the situation each of us is in, in our deeply unjust world. When our institutions fail to protect the vulnerable, we face endless conflicting choices. In the face of extreme suffering, there is always more I—or you—could do to help. There is more that, from the moral perspective, each of us has a fundamental reason to do. There is always someone else we could save. When we are daily upholding structures that sacrifice others, that moral pull is (or should be) stronger again. It is reasonable, even appropriate, to regret not doing more, even if I've already done all that could reasonably be expected of me.

I don't suggest for a moment that I have reached this point. But I still feel this two-way pull, always. Many of the activists I'll talk about in this book have reached it, and *they* still feel drawn to carry on.

And this is the situation which, as things stand, we bequeath to our children.

I've made this point philosophically, by drilling down on what it means to be a human. We can also make it empirically. Racism, says commentator and strategist Heather McGhee, "drains the pool." Quite literally: towns like Montgomery, Alabama, closed public swimming pools in the late 1950s rather than let Black children swim in them (as the courts said they must). White children cried too, as the bulldozers came in. Powerless and uncomprehending, they were stymied by an adult prejudice that cut off all their noses to spite the face of anti-racism. McGhee describes white Americans rejecting healthcare reforms that would help them, rather than share them with compatriots of color. That helps no one.[9]

"A divided society," says the *World Happiness Report*, "has a hard time providing the kind of public goods that would universally support each citizen's ability to live a happier life." Nordic countries, with well-functioning democracies, low crime, high levels of trust, and generous welfare states, consistently score highest in the report. In 2014, political scientists surveyed OECD data from 1981 to 2007, asking whether more generous welfare states and more extensive labor market interventions made citizens happier. Their conclusion? A resounding yes, *including for better-off citizens*.[10]

The same applies to responses to the pandemic, or antibiotic resistance. Far from being rational, antibiotic and vaccine nationalism are insanely short-sighted. Resistant microbes can cross borders, so no one is safe until we all are. That's not just what I think: it's one of several very blunt warnings from the UN's Food and Agriculture Organization. "If you want to outpace the pandemic," says Sophie Harman, global health politics professor, "you have to vaccinate everyone as fast as possible. The way I see it," she adds, "it's not an 'either or.' Vaccinate in the UK or the US and allow others to make their own vaccine. Take your foot off the necks of people in low-income countries."[11]

In other words? (McGhee's words, in fact.) This isn't a zero-sum game.

AGAINST MORAL SELF-INDULGENCE

As a child, Evelyn Adele went camping with her mostly Black Detroit Girl Scouts troop. She loved it: the toasted marshmallows, the forest, and the singing round campfires. Meanwhile, the mothers leading the troop stayed up all night. Not for fun, but to watch out for the Ku Klux Klan.[12]

Eighteen-year-old Shafiya Aktar was forced to the slums of Dhaka when her coastal home was swept away by rising seas. Six months pregnant, barely more than a child herself, married when she *was* a child, she told UNICEF that she dreamed of her baby growing up and getting an education. It will be hard enough just to keep that child alive.[13]

These experiences—experiences I cannot begin to understand, in my privileged life—are a whole different order of suffering from the moral and psychological marring I just described. They are first-hand atrocities, searing, lived-through tragedies, not the second-order pain of the person who cannot do enough to help.

Let's be clear, then, about what I am saying, and what I am not. Sparing our own kids this moral distress is *not* the main reason comfortably-off parents like me should challenge global poverty, racism, and injustice. It would be outrageous, disgusting, to suggest that it is. You and I should pursue a just world, first and foremost, because *we* are moral agents. I can say this unequivocally, as a moral philosopher. I said it, at length, in chapter 1. This is what we owe to women like Evelyn Adele and her mother; to Gianna Floyd, whose father George was murdered by a police officer; to the victims of pandemics and climate change; to refugees from Syria or Yemen or Ukraine, doomed to die like Alan Kurdi died. This is what it means to be a good global citizen.[14]

But what I have laid out is *yet another* reason why privileged parents mustn't ignore these claims. We can't set them aside on the basis that we are (we think) doing what's best for our kids, and it's legitimate to prioritize them. My kids are global citizens too, or will grow up to be so. They are moral beings. So are yours. So are Bob's. Our insularity wrongs them too.

Harriet Shugarman gets this. "It's a personal motivation," she says, "but it has to be for all of our children." Many young people get it too, even when their parents don't. Even when sections of the media choose to cut it out (quite literally, in 2019, when Ugandan activist Vanessa Nakate was cropped from a photo of climate activists). White teenagers join Black Lives Matter marches. Youth activists in the Global North demand climate *justice*, like Fuyani, not merely action to protect themselves. "Some of us are fleeing the war," said Ukrainian climate activist Ilyess El-Kortbi in March 2022. "Some can't because they are hiding from Russian bombs." His fellow activists held solidarity strikes.[15]

In August 2019, four hundred climate activists from thirty-eight countries signed the Fridays for Future Declaration of Lausanne. They had three demands: keep global warming below 1.5°C, "ensure climate justice and equity," and attend to the best "united science currently available." "We fight for the liberation of all people," declares the Sunrise Movement, in the United States. "We are fighting to become the generation that turns the tide against racism and the institutions built upon it."[16]

BEING PARENTS AND BEING ANCESTORS

If things look scary enough for our children, they look even more so for theirs. As deforestation continues, the risk of another pandemic escalates. The threat of antibiotic resistance is only getting worse. Unless emissions are rapidly reduced to what the Intergovernmental Panel on Climate Change calls "low" or "very low" scenarios, temperatures will *very likely* increase by at least 2.1°C by 2080–2100 (compared with pre-industrial levels). They could be 5.7°C higher. The world would be a terrifying place.[17]

Often, philosophers invoke the idea of "stewardship," a notion already widely appreciated in Indigenous ways of life. We inherit something valuable from past generations, and must play our part in this chain, by protecting it for those still to come. I find this plausible but can make my point even more simply. Future people should matter to us because they, like us, are human. As well as global citizens, we are *intergenerational* citizens. Does this mean we must leave future generations a standard of living at least as good as ours? Perhaps. But it undoubtedly means this: it is wrong to live in a way that makes future people sick, to leave them destitute or at the mercy of endless storms, just as it is with people who live now. Basic morality tells us that.[18]

As parents, we also have special ties to future people because they are our grandchildren, great-grandchildren, or even more distant descendants. "I'm very focused on being a good ancestor," climate scientist and communicator Sarah Myhre tells me, "because in a very short amount of time I will be my kid's ancestor and part of my family line."[19]

Let's think about what that means.

In May 2022, Giovanna Lewis and Annie Webster, both grandmothers, glued themselves to a table during a meeting of Dorset County Council. They called themselves "grannies for the future," and were responding to the Council voting to expand fossil fuels. Council leader Spencer Flower called them "anarchists."[20]

In 2019, two grandfathers, Peter Cole and Marko Stepanov, went on a twenty-five-day hunger strike in London. They called on the Conservative Government to shift its net zero emissions target from 2050 to 2025. Otto Cars, who founded the network ReAct to campaign for action on antibiotic resistance, is a grandfather and, he tells me, motivated as such.[21]

My parents-in-law come on climate protests with my girls and me. They craft placards. They march through Edinburgh, calling for the future they want for their grandchildren.

These grandparents love their grandchildren and are showing it. They are also acting as they should (or, in Cole and Stepanov's case, going beyond it).

Here's one philosophical explanation. We have special duties not only to our kids, but also to theirs, and even to our more distant descendants. In bringing my daughters into existence, their dad and I did more than create *them*: we created, potentially, two whole lines of future people, exposed to the vicissitudes of the world. Do I not, then, owe it *to those future people* to protect them as well, if that's something I can do? On the same reasoning, our (biological) parents owe (or owed) it to *our* kids to protect their future; a responsibility that grandparents like Lewis and Webster, Cole and Stepanov all recognize.[22]

Here's another philosophical explanation.

We are not islands. My daughters cuddle toys, each other, their friends and grandparents, and cats. They pour out love. That, too, is part of being human. If my girls live the lives they deserve as human beings, they will be able to live those lives with others, starting with me, their dad, and each other, with the aunts and uncles, grandparents and cousins who adore them. They will love and be loved, care and be cared for. They will keep friends they already have, and forge new friendships. They will have boyfriends or girlfriends, husbands, wives, or partners. If they want to, they may have children of their own.

This is wonderful, but it also exposes them to the same potential for suffering that having them has brought us. Truly to love another person is to make their well-being your well-being. The untimely loss, the serious illness of spouse, parent, or sibling, blights your life, too. Most of all, the suffering of a child does so. Even the prospect of such suffering, later in your child's life, is devastating. Parents, when dying young themselves, focus on their children. They fear for their children's future, write letters to them as young adults. A mother knowing her child must live in constant danger has her own life marred.[23]

In other words, if we owe it *to our own kids* to help their lives go well, we must protect *their* children's future.

We might go further, with philosopher and activist Rupert Read. He thinks the argument iterates. For our children to flourish, their children must be able to lead a decent life; for their children to do so, *their* children must also be able to do so, and so on. If we love our children (or, for that matter, our nieces and nephews, or friends' or cousins' kids) we must care about our (and their) distant descendants. Part of being a good parent is being a "good ancestor."[24]

This in turn, says Read, means being a good intergenerational citizen. He explains it like this. Our descendants are likely to disperse across the globe. That's what people do: they emigrate; they go abroad for study or work and meet partners there; they are driven from home by disaster and end up never coming back. Will my great grandkids all be here in Scotland, or even in Europe or the Global North? I doubt it. Of course, my daughters may not have kids, or theirs might not. But they might. So, Read says, if I truly care for my children, I must care about everyone, across the world, in many generations' time, and I must do what I can to protect them.

OUR CHILDREN AND OTHER ANIMALS

No man is an island, but neither is humanity itself. And as humans, or rather, as the comparatively rich ones, we are destroying much more than our own posterity. "Nature is unravelling and . . . our planet is flashing red warning signs of vital natural systems failure." That was Marco Lambertini, director general of WWF International, in 2020. The Living Planet Index, which measures biodiversity, fell by 68 percent between 1970 and 2016. By 2020, the species extinction rate was estimated to be 117 percent higher than normal (that's "normal" for the past two million years).[25]

This matters in its own right. I'm sure of that, as a moral philosopher who has written and read and published on this, too. Nonhumans have moral claims on us; we are *wrong* to do them wanton damage. Many other mammals, and some nonmammals, are "like us" in central and relevant ways. It's not just that they can suffer, although they can. They feel stress and fear and pain. They can also live incredibly complex lives. They can be sociable and affectionate, or experience fulfillment in parenthood. They have interests. They have, as philosopher Martha Nussbaum puts it, dignity.[26]

Species and ecosystems can also flourish or not flourish, in complex and beautiful ways. They have if not dignity, then what political theorist David Schlosberg calls "integrity." If a river system is undermined and its water polluted or if polar bears vanish altogether (as they're predicted to, by 2100), something irreplaceable has been lost.[27]

Even if you don't think the nonhuman world should be protected for its own sake, you must, I think, accept this: its destruction matters *instrumentally*. It matters for our children. It matters directly—emotionally, psychologically—and it matters fundamentally because it is an existential threat.

When I think of a time that it feels like I've got parenting nailed, it goes like this. Campsite, good friends gathered round. Kids climbing trees, filling wellies with stream water. Adults drinking red wine out of cups. Fresh air like a tonic. Owls calling at night. Laugh, breathe, sleep, repeat. And not a screen in sight.

Xoli Fuyani takes Black teenagers from Cape Town townships on hikes. "There's this sense of wonder," she says. "I love that marvel, looking at them looking at these tall trees they've never seen before. At the waterfalls." She describes their initial fear of the unknown, then a dropping of guardedness. A surrender. "It's so deep that I can't explain it, but it's also about being kids, running around and being free."

Even if our children or grandchildren could live in a bubble, feeding at a distance off the remnants of "nature," but entirely separate from nonhumans, that would not be a flourishing life.

In the Global North, we appreciate this inconsistently, even self-indulgently. A few years ago, my friend, who has at least a lively sense of his own inconsistency, told me that the £100 vet's bill to treat his son's pet rat was pretty much exactly what they paid a few years earlier, to get "pest" rats exterminated from their home. More generally, we get misty-eyed over landscapes that are, too often, the denuded remains of what they should be, because they have been stripped to make way for animals we can shoot, eat, or make clothes from. We form sentimental attachments to "cute" or funny or engaging species.

Those perceived as ugly or dull we largely ignore (unless we can cook them).[28]

But, still, interacting with nature matters to us. Study after study confirms that. It matters physically. A 2018 Finnish study created "forest floors" in the gardens of day care centers, and children's immune systems improved. It matters mentally. A 2019 survey of 450 UK children found that Wildlife Trust-led activities increased children's personal well-being. A 2018 review of observational studies concluded that green space helps with children's emotional and behavioral difficulties. According to the *World Happiness Report*, being in parks and allotments, or by ponds, lakes, canals, and rivers, makes city dwellers happier.[29]

We grieve, sometimes, for lost species. "Will he ever see a moose?" asks activist Naomi Klein of her young son. "Will he ever see a bat?" In 2016, conservation expert Terry Hughes tweeted, "I showed the results of aerial surveys of #bleaching on the #GreatBarrierReef to my students. And then we wept." In London the house sparrow, once prevalent, has all but vanished. The reasons are varied, but anthropogenic: lack of nesting sites, pollution, traffic noise. Researchers Helen Whale and Franklin Ginn interviewed citizens of this bustling, apparently hard-headed city, and found a sense almost of *more* than loss, as though the city was haunted, the spaces redefined, by the absence of this one little brown bird.[30]

In all this, we glimpse snippets of a wider truth—a truth that should, perhaps, be obvious from everything we saw in chapter 1. It is a truth that Indigenous cultures, once again, have appreciated much better than the rest of us. I do not want to leave my daughters a world denuded of its rich diversity. But if I properly recognized them for what they are—*part* of this vibrant world, not merely parasitic on it—it would be unthinkable that I could do so.[31]

Our children need the nonhuman world, as far more than the commodity it has been turned into. They need it because they (and their descendants) cannot survive without it. Trees and oceans regulate the systems that keep us alive. Food production, clear air, fresh water, medicines, all rely on complex, multispecies ecosystems. Biodiversity destruction and the mass mistreatment of nonhuman animals aren't merely "bad" for our children; they are destroying their future. The same patterns of oppression undermining women and people of color, for many generations, are the *same* processes, ecofeminists point out, that have exploited and destroyed the nonhuman world.[32]

Use of antibiotics that are "critically important" for human health went up 91 percent worldwide between 2000 and 2015, just among people. But nearly three times as many antimicrobials are given to farmed fish and animals, often preventatively. Agriculture drives 80 percent of deforestation, and cattle "replaced nearly twice as much forest as all other commodities combined" between 2001 and 2015. This makes climate change doubly worse: through methane emissions and the devastation of a carbon sink. It also increases the risk of future pandemics. Destroy biodiversity, say ecological modelers, and you get fewer species, but more animals within those species. Those species, too often, host pathogens that could make us very, very sick.[33]

As moral agents, we are also "ecological citizens," interconnected at this deeper level. This is true wherever we start morally: whether we focus on caring relationships (caring *for* nature, as well as our fellow humans), or the intrinsic value of systems from Gaia down, *or* the interests of individual humans or nonhumans. Whatever else being a good ecological citizen means, it means playing my part in redressing these patterns and relations that have got so very badly out of sync. And we can't be good parents without doing all this.[34]

Here's one more thing to bear in mind, before we get to what you (or I) can do about all this. The knowledge of *our* complicity could cause deep, additional pain to our children, and theirs.

"It felt like a punch in the gut." That's one Reddit user's comment on finding out his ancestors had owned slaves. "For the next ten years, it burrowed its way deep inside me," wrote journalist Martin Davidson, "until I could bear it no longer." He had learned that his grandfather was a Nazi. Unless you and I act decisively, our children, grandchildren, and great-grandchildren will know that their parents and grandparents were part of causing widespread suffering, *including to them*, and did not try to stop it.[35]

So let's talk about what we can and should do. There are still plenty of difficult choices, moral and psychological. But I'll start with the toughest question: Should we be *biological* parents at all?

To be good parents, we must be good global citizens, good ancestors, good intergenerational citizens, and good ecological citizens.

We must be good global citizens, pursuing a just world, because our children will be moral agents too. As such, the suffering of others matters to them. That shouldn't be our primary motivation for challenging fundamental injustice (it would be disgusting if it were!), but it strips us of one possible excuse for not doing so.

Good parents must be good ancestors and intergenerational citizens because our children may have children of their own. How well their lives go will depend on the prospects for those future people, even as our well-being depends on theirs. But those kids, too, may have children, and so may they, so truly protecting our own children means protecting distant descendants.

Good parents must be good ecological citizens because our kids need the nonhuman world. Plausibly, they need it directly: it is good for them (and us) to interact with other animals or be in green spaces. They undoubtedly need it to survive at all. Unless we stop exploiting and dominating the nonhumans who share this planet with us, we can't address the global crises that threaten our children's future.

II   WHAT WE DO NOW

4 THE HARDEST QUESTION

Should we be parents at all, at least biological parents? It's not an easy question to ask—or answer—but it's out there, so I'm going to take it seriously.

Begin with this: Why *do* we have kids? "Because I didn't know any better," quips one mother on the leading UK parenting forum, Mumsnet. "Pure accident," says another. "Met a nice chap, had a bit of a roll in the old hay and 'whoopsa daisy.' Turned out to be perfect for me."[1]

In the United States, 45 percent of pregnancies are unplanned, though that doesn't make them unwanted. But for many couples, the question of whether to be parents is more thought out. In 2005, psychologist Darren Langdridge identified the six reasons that most influenced couples to have kids. "Raising a child would be fulfilling." "I feel it would make us a family." "I would give a child a good home." "My partner would be pleased." "It would be something that is part of both of us." "Biological drive." His survey was limited to white, married, heterosexual couples in the UK, but a 2012 study found similar motivations among US gay men looking to adopt.[2]

Unfortunately, there are other reasons, too: reasons *not* to have a child, especially now. Put brutally, is having a child bad for the world? That's what Brandalyn Bickner, a Washingtonian who worked for the Peace Corps, thinks. She grew up in a big family, but in 2019 she told the BBC that she wasn't having kids herself because of the carbon price tag. Or is the world too bad for any potential child? In 2018, British musician Blythe Pepino started the "Birthstrike movement," for those who thought just that. "I really want a kid," she said. "I love my partner and I want a family with him, but I don't feel like this is a time that you can do that."[3]

Almost four in ten young people are hesitant about having children because of climate change. Birth rates dropped in the United States and Europe during the COVID-19 pandemic. As one woman put it, during the Trump presidency, "Even before the pandemic, we were worried about climate change and the impacts of the current presidential administration on the world. It feels like if you have a pro-con list for whether or not to have a kid, the con list is just growing." With the Russian-Ukraine war, that list just got longer.[4]

These aren't enjoyable thoughts, especially if (like me) you already have kids. But they are real concerns, so let's see how far they take us.

THE COST OF HAVING CHILDREN . . .

Here's an uncomfortable fact. Having a child in the United States increases carbon dioxide emissions by 7 tonnes per year and in France by 1.4 tonnes per year, compared with reductions of 0.4 tonnes per year for shifting to a plant-based diet, 2.4 for living car free, 1.6 for avoiding a transatlantic flight and 0.01 for reusing a plastic bag one thousand times. Among the life choices we make, this has one of the biggest carbon footprints. Put another way, in a society like mine, having a baby means creating someone who will use a disproportionate share of an already very limited ecological budget.[5]

Let's assume, for now, that part of being a good global citizen is cutting your carbon footprint. That amounts to a moral reason *not* to have children. But is it a decisive reason? Does it make you or me wrong to become parents at all, or to have already done so?

My short answer is no, but we should take these numbers seriously. The longer answer is complicated, so bear with me.

I want to begin by being very clear, since this could very quickly drift into morally dangerous terrain.

First, in talking about individual carbon footprints, I am *not* letting governments or corporations off the hook. We should be very alarmed by the policies or public narratives that push the onus for change onto individuals, especially individual women. Carbon-cutting is just one thing that some of us can and often should do, as part of the bigger picture. It's not the most important one. (If you're wondering why, the next few chapters will explain.)

Second, I'm talking about the individual choices of affluent would-be parents, mostly in countries like mine, who have a meaningful choice about whether to have kids and whose children would likely have a hefty carbon footprint. I'm not discussing population *policy*, which has, in practice, violated human rights in egregious and appalling ways. And it is racism, pure and simple, to use the "population question" to shift blame for climate change onto the global poor.[6]

You might still want to stop me here, even with these caveats. Children, you want to say, are not products or "lifestyle choices" to be chalked up to the carbon tab along with everything else. They are *people*.

They are, but that doesn't make the whole conversation a mistake.

Let's be clear on this, too, since it's morally foundational. *Anyone who comes into the world is as morally important as anyone else.* Any child deserves to flourish, to love and be loved, care and be cared for, and to build a life they can value. What's more, new people face crisis on crisis, including terrible mental health burdens, because of what others have done to their planet and their society. It would add insult to already-terrible injury to imply that some of them aren't entitled to their place on this earth. They are entitled to much more than we are likely to leave them.

But we *don't* imply that, in asking whether we should have biological children. There's a distinction between what philosophers call ex-post reasoning (reasoning after the fact) and ex-ante reasoning (before it). Just because every new human is valuable doesn't mean we must always bring more of these valuable beings into existence. (Go far enough down that line, and we reach what philosophers call the "Repugnant Conclusion": we must keep on adding more and more people, stopping only when their individual lives are so bad that they would no longer be worth living.[7])

As couples and individuals, we can have rational conversations about whether to have children without doubting for a moment how morally and personally precious those theoretical possibilities would become once they were real, breathing people. In fact, we have these conversations all the time. We mull over the financial implications of parenting, or the freedoms we'd give up. Parents don't love their children any the less for having thought carefully about whether to have them.

Perhaps, though, that wasn't your point. Children, you say, are not only people: they are different people from their parents and responsible for their own decisions. If your son or daughter contributes to climate change, that's on him or her, not you.[8]

In fact, it's not that simple. *Of course* as parents we are not responsible for every warped decision our children go on to make. The parents of a fossil fuel executive or the parents of a president who orders mass deforestation are not responsible for the suffering their child causes, solely because they produced that child, any more than the parents of a serial killer are. (They may be accountable for all sorts of other reasons, but not that one.) But this is different. It's not unusual to have an unsustainable carbon footprint in societies like ours; it's precisely what we can expect our kids to have, not least because it will be difficult for them to do anything else.

Am I responsible for the choices another person makes purely because I put them in a position where they have a free choice? Obviously not. But do I share responsibility for the harm they do if I put them in a situation where it is very difficult for them *not* to do it? I think so.[9]

But perhaps you meant something else again, by reminding me that children are people.

Kids are amazing. They are a gift, says climate communicator Sarah Myhre, talking to me from Seattle. They can transform the world in incredible and positive ways. Like me, she's a mother. Children, says educator and activist Harriet Shugarman, can be a force for change: "We don't know who is going to be that inspiration, that young person who invents that new technology." Yes, there are environmental downsides, to bringing a new polluter into the world. But there are upsides, too.

Kids are important collectively. Our society needs young people: it needs them to maintain institutions, pay taxes, keep us and our social services going when we are old. It needs them to become the next social justice leaders and science, technology, and healthcare innovators. And it needs young people emotionally. If you doubt that, then read P. D. James's *The Children of Men*, where there is universal infertility, no babies are being born, and each generation moves inexorably toward not only its own unnurtured end, but also that of humanity.[10]

This is a reason (one of many) why the costs of raising that next generation shouldn't fall *only* on parents. But it's not a decisive reason for you (or me) to have a biological child. That's because we start, as individuals, from where we are, and each of us knows that we are not in James's dystopia. Plenty of other people *will* have children, and while our economic and social institutions need *a* next generation to maintain them, it's not automatically better to make that generation bigger. There are also other ways to plug the "ageing society" gap in countries such as ours. Restructure the working environment so older people can contribute more (and recognize that they do already, through unpaid care for grandchildren). Welcome young immigrants.

Kids are also important *individually*. As well as being people in their own undoubted right, they are a gift to their immediate community. They bring their own joy to the family and friendship groups that they become part of. My children each add something unique and wonderful to the lives of their grandparents, aunts and uncles, and cousins, just as my niece and nephews do to mine. Most of all, they add immeasurably to the life of their dad and me.

This matters, not because we can (or should) put our hypothetical children into some "cost benefit" equation of probable joy and economic benefits on the one hand, probable environmental cost on the other. It matters because demandingness matters. We saw that in chapter 2. Having a moral reason to do something doesn't always mean we *must* do it, if that would be very painful for us, or those around us.

It's still up for debate how much we must each do, in pursuit of a better world, although philosophers use ballpark terms such as "significant cost" or "adequately protecting valuable pursuits," reminding us that most of us do far less than we should. It's even harder to say how much effort and pain we should take on to cut carbon emissions from our own lives right now, because that's only one bit of what we should be doing. But I can say this. If you think young men and women should stay childless, because of the carbon footprint their children would have, you are setting the demandingness bar very high.

. . . AND THE COST OF *NOT* HAVING THEM

I cried a lot when my children were babies. I cried because *they* cried and couldn't tell me why. I cried from sheer exhaustion. I cry now, too, when things go wrong for them, or they have a day of being absolute horrors, or I feel I'm messing up at parenting even more than usual. But I laugh too, and I laughed then. I feel fulfilled and desperately anxious, delighted and infuriated, exhausted and full of life. And, sometimes, I am so overwhelmed with the quiet joy of being their mother that it is as much as I can do to breathe.

Having kids doesn't always make us happier. "It is difficult," says care ethicist Sara Ruddick, "when writing about motherhood—or experiencing it—to be balanced about both its grim and its satisfying aspects." Looking at sociological studies, the evidence is mixed. Reported happiness doesn't increase with having kids, or does so only in certain circumstances—for the first child, perhaps, or for married parents, or working fathers. But parenting, as philosopher Ingrid Robeyns puts it, is special. Our kids need a close relationship to thrive. We saw that already. That relationship can also be incredibly important *for us*.[11]

Parenting is about love: the spontaneous, unconditional, and unselfconscious love of a child for her parent, and a depth of feeling, in return, that is often unmatchable. Shakespeare tells us: "Love is not love / Which alters when it alteration finds, Or bends with the remover to remove," and this is truer in practice of parents than of lovers. Raising a child, say philosophers Harry Brighouse and Adam Swift, combines this deep love with intimacy and fiduciary care (making decisions on that child's behalf).[12]

The care, too, is of a peculiarly rewarding kind. Whenever we look after a living being, we have the experience of being responsible for their well-being. My cat, squeezing herself between me and the keyboard as I write this, shares a life with me. I feed her and care for her, decide the limits of her freedom. We care for other people we love, too: elderly parents, disabled family members. But it's different with kids, because all being well, we oversee the very process by which they grow out of that day-to-day care (though not out of our love!).

If philosophers like Brighouse and Swift are right, parenting is a unique human experience for which nothing else can substitute. Even if they aren't, it is still *one* incredible relationship, and a central life project for many people. That makes *not* doing it, if you want to, very hard indeed.

I'm not speaking from a neutral position. I know that. I must not generalize from my own certainty that my life would be fundamentally worse if it hadn't had my brave, kind, loving and lovable girls in it. Very many people have brilliant, full lives without being parents. "Not for a millisecond, no," the feminist Gloria Steinem said, when a journalist asked if she regretted not having had a child. "I have not become a mother, just as I have not become an acrobat or a brain surgeon," writes *Guardian* theater critic Arifa Akbar. "It might have been one kind of life and this is another: meaningful, rewarding, joyous."[13]

But for some, the need for a child can be almost visceral. I see this unscientifically, dipping in and out of blogs and forums, talking to friends, remembering my early thirties. "The ache gnawed at me," says a mother on one parenting forum. "From the time I was a young child," says another, "there's never been anything else I wanted more for my life. They're the whole *point*."[14]

We can't know exactly what the emotional or mental health toll would be of not having a child you desperately want, for environmental reasons. But a 2003 study found "substantial and significant long-term psychological distress" among women who had never become mothers because of infertility. Even their identities were threatened. Lizzie Lowrie, a British woman who endured six miscarriages, described not being able to have children as "the most difficult experience of my life. I have come a long way," she told *The Guardian*. "I know now that I don't need a child to have a life of purpose, but the desire to have a child—that never goes away."[15]

"Men experience a different kind of grief," said Michael Hughes, a fifty-four-year-old Australian, speaking out to his local radio station after he and his wife had spent years on unsuccessful fertility treatments (and hundreds of hours in what he called "lifesaving" therapy), "and we do it alone."[16]

PARENT, OR BIOLOGICAL PARENT?

Here's a thought that has probably occurred to you already. Yes, not being able to be a parent can be searingly awful, but you can parent without bringing a new person into the world.

We might think innovatively about what "family" means. The philosopher Anca Gheaus suggests that we explore ways of co-parenting with more than two parents, so that three or four adults can experience this mutually valuable bond with the same children. This happens already, to some extent, in blended families, or when a same-sex couple raises their child together with the other biological parent of that child.[17]

Or there's adoption. Adoptive parents are real parents; to suggest otherwise would be outrageous and insulting. Parenting can be incredible, loving, and mutually rewarding with a child who does not share your gene pool; adopting is a worthwhile and valuable thing to do. Yes, this too may increase carbon footprints (if adopters are relatively rich), but it does so by saving an *existing* child from serious deprivation. This weighs differently in the moral balance from bringing a new person into existence—so much so, in fact, that some philosophers think there's a *duty* to adopt, if you want kids at all.[18]

However, there are also reasons for wanting a biological, procreative child, and these can't simply be dismissed. Biological drive was one motive influencing couples, especially women, in psychologist Darren Langdridge's study. Another was the genetic link. As the study put it: "It [a child] would be something that is part of both of us." At one level, it might seem selfish: this longing for a taste of immortality. At another, it seems deeply human. It recognizes, as the philosopher Ingrid Robeyns puts it, that we are a "procreating species." Over generations, we have developed patterns and practices that find meaning in this handing down of genes, the creation *and* raising of new people.[19]

I feel this joy in continuity: not so much in reproducing part of myself as in passing on the traits of those before me, whom I have loved. I feel it reading my mum's expressions in my older daughter's lively face, or seeing my much-missed grandmother in my little girl's blue eyes and gentle stubbornness. It is a deep and abiding gladness that something of my grandma persists in this wonderful new person, whom she never met.

Then there's the actual experience—physical and emotional—of carrying, birthing, and breastfeeding a baby.

I discussed this with a friend, walking across Edinburgh hills, mid-lockdown. We had both disliked being pregnant. We were exhausted and unremittingly sick. Labor had been an agonizing, two-day ordeal for me. My friend had birthed one of her babies in the back of a car: hardly a stress-free experience! But if either of us could have had our children—the *same* children—grown for us in a lab instead, or carried by a surrogate, we wouldn't have done it.[20]

We're not alone. In 1999, Rosemary Mann and colleagues, interviewing African American midwives, found that pregnancy was seen as a key transition in a woman's life. In the words of one participant, carrying a child is "specifically about being an adult and being a woman." Feminist philosopher Amy Mullins thinks we should "reconceive pregnancy" (pun intended). We should recognize it as an experience in itself, positive *and* negative, some elements of which are unlike anything else, and not just as a biological process.[21]

Does that mean no one with a womb can enjoy a full life without these experiences? Obviously not, since very many do. If you can't relate to any of these reasons for wanting a biological, procreative child, but being a parent is important to you, maybe you should adopt instead. I'm just saying this: for some, possibly many, going without these intense and unique experiences would be a very serious loss.

This isn't the end of the conversation. For one thing, becoming a parent doesn't necessarily mean having more than one child. For another, there's a still more uncomfortable possibility to consider. In giving ourselves that incredible experience, in bringing children into this world full of fear, do we wrong *them?* I'll talk about that next.

A WORLD NOT WORTH INHERITING

"There's scientific consensus that the lives of children are going to be very difficult. And it does lead young people to have a legitimate question: Is it OK to still have children?" So said Congresswoman Alexandria Ocasio-Cortez in 2019.[22]

"I've always known that I've wanted to experience pregnancy, birth and motherhood. But now that the planet's future is so bleak and our leaders are so unwilling to act, it seems as though it would be incredibly selfish and morally reprehensible for me to decide to bring a person into a world [where] I know they likely won't be able to live a full life in." Those were the words of a woman identified only as "Ashley," speaking to *HuffPost* in 2019.[23]

In 2020, researcher Linden Ashcroft was expecting her first child, and writing an open letter to the website Is This How You Feel?, which collects scientists' emotional responses to climate change. "Lately," she said, "I mainly feel guilt, and grief . . . I am acutely aware of how selfish it is to bring a new person into a future that looks so grim. I feel guilty for doing that to her, and for partially causing that grief."[24]

These women's reasoning, like Blythe Pepino's, is not reducible to carbon counting. They fear that any child they have will inherit a poisoned chalice: a devastated and dangerous earth.

Sarah Myhre gives this line of argument short shrift. She reminds me: "There have been many people and cultures bearing witness to moments of genocide and removal from their lands of their own heritage." (As Mary Annaïse Heglar puts it, the United States has been an existential threat to Black people for centuries.) The anguish, Myhre says, is genuine, but the decision itself comes from a position of privilege. "White middle-class women are not having babies because they believe they were entitled to a life that was going to be easy and was going to continue to have wrap around services for all the hardships they would be protected from."[25]

As a middle-class white woman, I would do well to remember that. I must recognize, too, the privilege inherent in being able to make this choice, as a woman, and to have it respected. As a philosopher, I can also question the underlying argument.

My children's lives, and yours, will be made harder and more dangerous by these global crises. The world we leave them will not be *good enough*, because it is not what they deserve as human beings and global citizens. But that does not make those lives irredeemably terrible, or not worth living. Their futures are clouded by the risk of disease and extreme weather, insecurity, injustice, and all the mental health burdens that go with that. However, they can also experience joy and care. They can fall in love, read books they want to lose themselves in, and laugh with friends.

Here are two points to consider. First: If our kids' lives would be rock-bottom terrible, the discussion we just had about demandingness and reasonable sacrifice would no longer apply. Recall the core moral principle taken for granted in chapter 1: don't do serious, avoidable harm. That's undoubtedly extremely demanding: a straightforward ban on certain individual actions. If we would be fundamentally harming any new child by bringing them into the world, then we *shouldn't* have one (if we have a choice), no matter now painful that is. But is it necessarily wrong to have a child because they will likely miss out on some central human experiences, even if they will still have a life worth living? That seems less clear cut.

Second: The future isn't fixed. *We can still change it*, for better or worse. The choice young people face is not this: (1) Have kids, and leave them a doomed world *or* (2) don't have any kids at all. There's a third option: Have kids, and cooperate with other adults to leave them a thriving, biodiverse, and socially and environmentally *just* future. In other words, children can still have the lives they deserve, if norms can be challenged, institutions reformed, and governments and corporations held to account.

That said, there is still a lesson to be learned from this very real pain. Even if we're not yet in the tragic situation of not being able to have a child without their life being utterly terrible, our kids could be in that situation, or their kids could be.

When my little one was four or five, she spent much of her time either demanding a little sister or telling me that she wanted to "born a baby" herself when she was grown up. That made me smile. It also made me want to cry.

If my daughters want to become parents when they are older, I don't want them to have to give that up—that complex, demanding, infinitely rewarding relationship that *I* have with *them*—because things are so bad by that point that they would wrong any child they brought into the world. We can put this more strongly. If our children are in this situation, then they too have been terribly wronged. And hard as it is to admit it, I know this as well: if I hadn't allowed myself this incredible bond, I wouldn't have bequeathed that potentially tragic dilemma. Having done so, I owe it to my children to do all I can to make sure they *don't* face it, in their turn.[26]

Or here's another line of argument, recalling something we talked about in chapter 3. Even if their kids would have lives worth living, the situation we are leaving our kids is hardly an enviable one: their children will still face huge risks. The philosopher Anca Gheaus formalizes this. Because parenting well can be a uniquely valuable experience, she says, we all have an interest-based right in being able to raise children we *can* parent well, or at least adequately. That is, our children are entitled to have *and* raise children who have flourishing lives, not lives that are merely worth living.[27]

Like Rupert Read (who I also discussed in chapter 3), she goes further, iterating the argument. If my *grandchild* has no chance to parent adequately, my child's (their parent's) life is also not fully flourishing, because they cannot enable them to enjoy this key human interest. And so on. Ultimately, then, I would not be adequately parenting, either. But *we* can still turn this around, by making it possible for future generations to lead decent lives.

WHY WE (STILL) NEED TO TALK ABOUT FAMILY SIZE

"I have one child," wrote environmentalist Bill McKibben in 1998. "She is the light of my life; she makes me care far more about the future than I used to. And I have one child; so even after my work I have some time left, and some money left, and some energy left, to do other things."[28]

The question isn't just whether to have kids; it's how many to have. One reason for a small family (at least of biological kids) is the carbon price tag. Another is what McKibben alludes to: the massive to-do list we already have, if we want to be good parents and good human beings. If taking on more projects or commitments would mean doing less for those we are already morally bound to help, then that's a moral reason *not* to take them on.[29]

Parenthood, says philosopher Sarah Conly, is not like being a "Better Boy tomato plant," producing as many offspring as possible. What matters is the parent-child relationship. Important as the "pregnancy experience" may be, says ethicist Tina Rulli, we don't need to do it over and over again. Their point, broadly, is the same. This is not a numbers game. The implication? For Conly and Rulli, that one biological child is enough. McKibben puts it more mildly, in the title of his book: *Maybe One*. Some other thinkers settle on "at most" two per couple (one per parent). This has intuitive appeal as a "replacement rate": you and your partner are just reproducing yourselves, not increasing the number of people per generation. It also gives your child the experience and support of a sibling.[30]

I don't have a decisive figure. I don't have one because, apparently compelling as the idea of the replacement rate is, whether two children per couple *is* "sustainable" (or, for that matter, whether one child is) depends on what happens collectively. In other words, it depends on whether and how much government policies or changing norms cut everyone's environmental footprint.

I also don't have a universal answer because, ultimately, I'm not sure there is one.

Stopping at one child, or two (or three or four, for that matter) is much more demanding for some people than for others. Of course, never being able to experience birth, pregnancy, or parenthood, if you want to, is on a different scale of awfulness from not being able to do it repeatedly (or indeed from *having* to do it repeatedly, if you don't want to). However, it can be very distressing not to have more biological children. If you doubt this, read academic and writer Pragya Agarwal's painful account of her own battle with secondary infertility, or consider the health risks that women run to have second or even third or fourth kids. What's more, these differences can result from cultural or religious views that are outside individuals' complete control. They can even be a legacy of historic atrocities.[31]

For generations, Black and Indigenous women were either forced to have children in unthinkable conditions or deprived of the right to have them at all. They endured slavery, eugenics, and sterilization without proper consent, as Jade Sasser, gender studies professor, reminds us. No wonder, she says, that it matters fundamentally to some women of color to have and raise large families of their own biological kids, with the partner of their choice. It would be more than outrageous for a white woman like me, unscarred by any such cultural trauma, the beneficiary of centuries of fossil fuel emissions and systematic oppression of women of color, to pass any comment on that. It would be unforgivable.[32]

With all this in mind, I will speak only for myself, as you must decide for yourself. I have two children. That's how my husband and I struck the balance between our own wishes and our commitment to climate justice.

It came at a cost. We both would have liked at least one more, and I wanted four kids. I don't think I realized how much I wanted that, until we were deciding *not* to try for a third. For months, I daydreamed about that alternative world: about a dream toddler on a balance bike, chasing after their sisters; another baby in a sling. But I know how lucky I am. I've had those nothing-like-it early years, with both of my babies. I've experienced (if not exactly *enjoyed*) pregnancy and childbirth. I have that wonderful-and-petrifying, life-enhancing relationship with my girls, and I can witness theirs—tempestuous, absorbing, and deeply loving—with each other.

For me, in the world as it is, that is enough. I wish previous generations, and years of political failure, hadn't left me with this choice. But we are where we are (and I am far luckier than many others). Getting it right for my amazing children will be time consuming enough, without having more of them.

So let's talk about how I can do that.

Chapter 4 in Essence

This is the hardest question to answer: Should we be biological parents at all?

One reason not to have a child, in high-living, heavily polluting countries, is the carbon footprint each of our children will have. But eliminating this by forgoing parenthood is too much to ask of many of us, because parenting is an incredible, perhaps unique, human experience. And for many, it would be a serious loss not to do this. There are plenty of moral reasons to adopt a child instead, but it can also be a major sacrifice not to be a biological parent. As humans, we see ourselves as a procreating species. Some of us even find a sense of immortality in passing down our own genes, or those of parents and grandparents we love. Many women value pregnancy, birth, and breastfeeding.

Another reason not to have a child is that the world is too terrible to bring anyone new into it. But our kids' lives can still be worth living, even if they fall short of the "decent life" they are owed. Moreover, this is a false dichotomy. Rather than have no children, or doom them to an unjust and diminishing world, we can have them, and build a better future.

All that said, there are moral reasons to think carefully about family size. It matters for carbon footprint reasons. It also matters because having more and more kids might just mean not doing enough, for the ones we have already—or for everyone else.

5 BEING ACTIVISTS

It's a phenomenon as old as time: parents' love for their children, and their urge to protect them.

It was there in the mothers of disabled children in Canada in the 1950s and 1960s who demanded community-oriented services and meaningful help for their kids. Their efforts launched what would later become a national organization. It drove a group of Black mothers in Texarkana, who learned in the 1980s that they were raising families in homes built on toxic waste. Led by Patsy Oliver, they protested until, years later, Congress organized a buyout of the homes on their poisoned street. It was there in the parents and kids who lined the streets in the UK, Cyprus, the Netherlands, Spain, the Czech Republic, and Australia on International Mother's Day 2019, all demanding action on climate change. More than two thousand marched through London, with eleven eleven-year-olds at the front: a living reminder of the narrow window then left to avoid dangerous warming.[1]

Now that window has shrunk even more.

Sometimes, when I wake in the night, I am almost paralyzed by the enormity of how much is at stake for my girls' generation, and by my own smallness in comparison. I look for inspiration, and I find it in stories like those. This chapter will explain why.

FUTURE-PROOFING THE WORLD

Here's a nice idea: we innovate our way out of antibiotic resistance, climate change, and pandemics. Yes, things are bad, we reason, but those scientists are so clever, they are bound to come up with something new. This, alas, is a fiction.

Technology matters, crucially, if we are to leave our kids with the world they deserve. This includes adaptation technology, energy efficiency technology, and technology to harvest renewable energy. New antibiotics, vaccines, and other biotech also matter, as do lab-grown or otherwise "fake" meats. However, technology alone isn't enough. For one thing, enough people must *want* to use it, and be able to afford it. "To get renewables on track with net zero by 2050," warned a 2021 International Energy Agency report, "governments not only need to address current policy and implementation challenges but also increase ambition for all renewable energy uses." Previous technical transitions have taken far too long.[2]

Instead, we need to do something as simply put as it is daunting. We need to live differently. In the Global North, we need to stop treating the nonhuman world, *or* our fellow humans, as expendable, exploitable commodities. To do all this—and even to get essential innovation off the ground in the first place—we need collective will. We need political change, institutional change, and norm change.

There are three strands to tackling antimicrobial resistance: innovation to develop new antibiotics; improved access to existing drugs for those who need them; and stopping *un*necessary use, by humans or farmed animals. We can't address this or pandemic risk, says Sophie Harman, without basic primary healthcare across the world. "And science won't solve that. It takes politics to solve that."[3]

It takes politics to secure reparations for the centuries of harm that climate change exacerbates: adaptation aid for the Global South; climate finance for loss and damage; welcome and support for the millions of refugees whose lives are overturned. As I write this, the UK government plans to pack desperate immigrants onto a one-way flight to Rwanda. The glossy press conference does nothing to conceal the moral shame. Just as it takes politics, not "thoughts and prayers," to tackle a gun culture that leaves children in danger in their classrooms, so it takes politics for governments to fix emissions-cuts targets that will avert calamitous climate change, and to stick to them.[4]

The twenty biggest fossil fuel companies caused 35 percent of carbon dioxide and methane emissions between 1965 and 2018. Half of the biggest tropical timber companies hadn't even made a public commitment to protect biodiversity by 2020. It takes corporate reform to change that, to shift food and drink giants toward plant-based products or clothing manufacturers away from child labor, or to stop big business from producing mind-numbing levels of greenhouse gases through their cash and investments. Even companies like Google, seriously working on *reducing* the carbon footprint of their own products, still finance fossil fuels.[5]

It takes political change to make all this happen, starting with putting a stop to the hundreds of billions of dollars that governments (unthinkably, atrociously) pour into subsidizing fossil fuels each year, and the "perverse agricultural subsidies" of almost every country.[6]

It takes institutional change to end embedded injustices. It takes institutional change within states, from reforming the school uniform policies that amount, according to a UK cabinet advisor, to "victim blaming from a young age," to increasing rape conviction rates; from ending links between law enforcers and white supremacists, to stopping the disproportionate crackdowns on Black and brown protestors. And, perhaps, it takes global institutional change.[7]

Imagine this. A group of citizens from around the world, genuinely representative, coming together to deliberate on the future of human beings, and of the planet. Imagine them setting baselines: the needs of the vulnerable, of future generations, animals, ecosystems. Imagine them choosing principles to guide global decision making. Imagine they were listened to. That's what a group of institutions, scientists, and social movements have tried to achieve: a small-scale Global Assembly so far, with one hundred participants, but intending to get bigger. That, or something like it, is what the philosopher Stephen Gardiner thinks we need: a new constitutional convention, redesigning the global system to protect future generations.[8]

Then there are the norms that determine how many of us live. These, too, need to change. Fast driving, multiple overseas holidays, high-meat norms, throwaway consumerist norms.

"Stuff is just stuff," says Leanne Brummell, Australian activist and mother to a teenager. "It's not so important. People are important. The climate is important. Plants and animals are important." She's obviously right. But we don't live as though she were.[9]

From 1950 to 2015, 8.3 billion tonnes of plastics had been produced. Only 30 percent was still being used in 2015. More than half went straight to landfill or was discarded. 8 percent was incinerated. Only 6 percent was recycled and only a fifth of that was still in use in 2015: the rest had been (yes, you've guessed it) incinerated, discarded, or sent to landfill. This is partly a matter of unnecessary packaging. But only partly. It's also about the goods themselves: toys that cost far more than they are worth and end up in landfill anyway, clothes too often made in sweatshops.[10]

Those are the obvious norms. But there are also other destructive norms, which may not be at the front of your mind.

Consider another true story. As my sister was leaving the hospital with her second baby boy, a midwife told her, "You'll be back to have a girl." My sister would have been equally happy with a son or daughter and was already besotted with her baby. She replied with a fervent, "No. I won't." She hasn't been back. But anecdotes like this are not a one-off.

Women who decide *not* to have children are still stigmatized. They get called selfish when they may be precisely the opposite. But social norms go further than that, in their determination to circumscribe women's lives. They dictate not only that we should want to have kids, but also that we should want to have "one of each," and keep "trying" until we do. There are variants, of course: "You must want a girl." "Your husband must want a son." Etcetera. These norms are restrictive. They are sexist, as well as acting to dismiss or negate the trans or nonbinary communities. If the last chapter is right, they are also bad for climate change. Challenging them would be a positive first step, in many ways.[11]

Or consider this. In our world of misinformation, sustainability consultant Mike Berners-Lee thinks we need to cultivate a "culture of truth." In other words, deliberate lying needs to be a "career-stopping" mistake. Professional credibility should depend on being candid and open in argument, acknowledging every valid side, not stirring up controversy for the sake of it, and giving the clearest view of the evidence. In other words, we need the exact opposite of organized denial.[12]

PARENT POWER

As collective "to-do" lists go, they don't get much bigger than this. No wonder we retreat, instead, to the innumerable day-to-day lists we've already got stuck to the fridge.

I understand this. I feel it. But then I speak to Xoli Fuyani, who has now added a role at the global network Our Kids' Climate to her already formidable repertoire. I speak to Brummell, who cofounded Australian Parents for Climate Action and represents Parents For Future Global. She protests against the gas industry in her Queensland community, then gets up at 5:00 a.m. to talk to her fellow activists from around the world.

I speak to Maya Mailer, who has three kids, codirects Our Kids' Climate, and cofounded Mothers Rise Up.

"As parents, we can have a really loud voice," Brummell tells me, sounding both cheerful and determined across the thousands of miles between her Australian evening and my Edinburgh morning. She's right. We have what should be a direct route to the "hearts and minds" of corporate and political leaders: many of them are parents too.

In May 2022, parents from Sunrise Kids NYC paid their second visit to the Westchester home of Larry Fink, BlackRock CEO. They took their young children, dressed as superheroes, along with handmade cards and pictures asking him to "be a climate hero," support the Paris Agreement, and stop supporting oil, coal, and gas expansion.[13]

On Father's Day 2021, parents and children protested outside the offices of Lloyd's of London. Working alongside the global "Insure Our Future" campaign, Mothers Rise Up, Parents For Future UK, and parent-climate groups from around the world sent cards and gifts to Bruce Carnegie-Brown, chair of Lloyd's and father of four. They asked him to stop insuring some of the world's most damaging fossil fuel projects. Months later, Mothers Rise Up gave him a handmade advent calendar. They've secured meetings with the top brass, including with Carnegie-Brown himself, a rare feat for any campaigning group.

"Most people pride themselves on being good parents," Mailer tells me, "and we are saying to Bruce Carnegie-Brown, 'If you're a good dad that's inseparable from the actions you are taking at Lloyd's.' We're appealing to him not as a corporate leader, but first and foremost as a father. They said it's made him think. I'm not sure how another campaigning group could do it in that way."[14]

Of course, the proof of the pudding is in the eating, or rather, in BlackRock, Lloyd's of London, and other corporate giants dissociating themselves from fossil fuels. After they met Carnegie-Brown in May 2022, the mothers "left frustrated and disappointed." But they were also "more determined than ever." On June 13 of that year, they organized a song-and-dance protest outside Lloyd's of London, complete with kites, dancers in suits, and one in a Carnegie-Brown mask. It was based on the moment in *Mary Poppins* where work-obsessed father George Banks realizes that his children are what really matter.[15]

Our other not-so secret weapon? There are an awful lot of parents. That gives us economic power. The global diaper industry is worth more than $72 billion. The baby food market looks set to tip $96 billion by 2027 and the children's clothes market $325 billion. Then there are the buggies, bikes, and scooters, holidays, family cars, bus or rail tickets. The nurseries and schools and colleges. The phones and tablets and trips to the bowling alley. Many of us have pensions; most of us have bank accounts: money that we can demand is put to good use, rather than invested in collective destruction.[16]

Our numbers also give us political power, if only enough moms and dads would follow the lead of groups like Parents For Future or Mothers Rise Up. According to political scientists Erica Chenoweth and Maria Stephan, it takes 3.5 percent of the population to be actively involved in a peaceful movement for change, for that movement to succeed. In the United States in 2020, there were 63.1 million parents with kids under eighteen living with them. That's 19 percent of the population, and it's still nothing like *all* parents.[17]

Parents are already good at cooperating. "We're plugged into the school networks, communities, workplaces, boardrooms," Mailer points out, "and that's the power of being a parent, especially a mother. You create these networks in day-to-day life that you can use to carry this through." We have global networks too, or can build them, in this age of social media and international communication. Our Kids' Climate holds monthly strategy calls with parents from fifteen countries.

"We are working with mums in Australia and India, parents in Poland and organizers across Africa who feel the same way," wrote Mailer, after her "frustrating" meeting with Carnegie-Brown. "The pressure will continue to grow."[18]

WHAT ABOUT ME?

So much we can do. But where, and why, do I fit in?

Here's where I think we are, after the first half of this book. We have what philosophers call *shared duties*, to build a better world together. They're not the only moral obligations we have. We have a clear-cut duty not to hurt other people, directly, ourselves. We have special duties to our husbands, wives, parents, siblings, friends, colleagues, and other things we must do for our own children. But these shared obligations are deeply important. We have them as parents, as potential ancestors, as global and intergenerational, and even ecological, citizens.

As individuals, we must help fulfill them, as best we can. Yes, it's complex. This isn't like tucking my daughters into bed. It's not even like reaching out for others' hands, on a beach, calling to those around me to form a human chain, as they did at Durdle Door. But if parents like you and me do nothing, the shared duty—to our own kids, as well as to other people—is unfulfillable.

I've focused on duties, but we could come to a similar conclusion by thinking instead as virtue ethicists do, about the kind of people we want to be. What does "benevolence" mean, when those in need are countless and the roots of their plight are irredeemably global? What does "humility" or "temperance" mean when global resources are running out? Do we need new virtues, for these troubled times? These could be respect for nature, treading lightly on the world, or "mindfulness" of where the goods we use come from and what happens to what we throw away.

These are promising. But even with this particular philosophical hat on, I cannot consider myself entirely in isolation. Nor can you. That's because we *can't* be fully virtuous in modern society. What would it take to detach myself and my family completely from these convoluted and environmental harms? Even if I could do it (spoiler alert: I can't), we would be cut off so entirely from other people that we couldn't exercise other virtues. So we each need a tendency to reform, or work for political change. We need "practical wisdom," or what philosopher Zev Trachtenberg calls "judgement," to find a path through this moral maze.[19]

Here's a natural worry. *We* can do a lot, but my being part of it will make no difference either way. Virtue theorists might circumvent this, by focusing on character traits (although I still need to figure out what exactly "judgment" means). Other philosophers have several answers to it.

There is a small chance that what you will do will make a very big difference. "Unknowns" *have* started movements. Nobody had heard of Greta Thunberg when she first sat outside the Swedish parliament with her "School Strike for Climate" sign, any more than they had heard of the parents organizing the 2019 marches. When Stacey Abrams and Alexandria Ocasio-Cortez chose to become politicians, they couldn't know how influential they would be. Even if I don't start a movement, I could be the extra vote that swings the election, the extra person at the march that gets public opinion on the movement's side, or that one last caller to a congresswoman who gets her to pay attention.[20]

Or I can think like this. Even if I don't make a difference alone, it matters that I help do so. The outcomes of these collective campaigns aren't already determined, for better or worse. Whether we're talking fossil fuel divestment or changing a sexist policy, there are lots of people, like you and me, who are still undecided on how to act. *We* could make a huge difference between us. My contribution, if not strictly logically necessary, is what philosopher Julia Nefsky calls "non-superfluous."[21]

Or here's another widely shared moral concept, one we try to teach our kids from the first moment they reach for the same toy as their friend: fairness. Others are already putting in the effort to do what we must, morally speaking, *together*. People like Mailer, Shugarman, Brummell, Fuyani, and many other already exhausted parents. If I don't do my bit, then, as philosopher Garrett Cullity puts it, I'm free riding on their hard work.[22]

TOO MANY OPTIONS

I can do a lot, as part of this endeavor.

I can give money, or at least invest it wisely. If I do nothing else, I can and should do this. I can donate to nongovernmental organizations (NGOs) promoting climate justice, racial justice, domestic and international security, and gender equality. And I can donate to political parties that do the same; to youth organizers such as the Sunset Movement; to groups promoting safe antibiotic use or campaigning against deforestation; to ReAct, the People's Vaccine campaign; the legal funds of activists arrested for civil disobedience; or the families of those attacked by the police.

I can invest in innovation and technology: antibiotics, renewables and climate adaptation technology, even vegan meat substitutes. I can petition my bank or pension fund to do so, and not to put money into arms or fossil fuels. The carbon footprint of the average pension is twenty-six tonnes, according to a recent UK study. The world's sixty biggest banks invested $3.8 trillion in fossil fuels, in the four years *after* the Paris Agreement on Climate Change. If they won't budge, I could look at switching to another provider. Preferably, I would do this at the same time as lots of other people and make a big deal about it.[23]

I can give time. I can join groups like Our Kids' Climate, Parents For Future, or Moms Clean Air Force. I can join climate strikes, Black Lives Matter protests, the March for Science, women's vigils like the one held for Sarah Everard in London. I can organize groups and strikes and marches. I can start petitions or sign them. I can talk to local councilors and school authorities. I can campaign for progressive political candidates. I can vote, and campaign for a lower voting age. I can support initiatives like the Global Assembly or Scotland's Children's Parliament, injecting the voices of young people and children into the politics that determine their future.[24]

I can write to politicians, phone them, tweet them. I can demand legal reform to make harmful institutions take responsibility. I can ask my representative to support a Green New Deal and the TRIPS waiver, which makes it easier for countries in the Global South to produce the COVID-19 vaccine. I can engage in civil disobedience like Just Stop Oil activists in the UK or pipeline protesters in the United States. In other words, where my government has fundamentally failed its citizens, I can use principled, public, deliberate law-breaking to bring about change. This, too has a long philosophical history.[25]

In my academic work, I have developed what I call a "cooperative promotional" account, to help decide *which* of all these things to do. I must think about all the other people who are motivated to be good parents or good global citizens, figure out how *we* can best move things in the right direction, and decide where *I* fit within that.[26]

This means a lot. It means, probably, committing myself to activism on one or two of these vast injustices, since no one can do everything. As they are deeply interlinked, it also means *not* addressing any one of them in a way that makes the others worse. (A classic example? "Conservation colonialism," or driving Indigenous communities out of their homelands in the name of protecting forests.)[27]

It means considering which groups and movements are likely to make a difference. It means asking where I best fit, in all these, and what influence and resources I already have, or could acquire. For politicians, or corporate bigwigs like Carnegie-Brown, it means using that power as a force for change. For the very rich, it could mean imitating billionaire Chuck Feeney, and giving most of your fortune away. But *how* that money is made, and how it's invested in the meantime, matters too. For the rest of us, it comes down to asking ourselves what we are good at. It means being flexible, driven by what really matters but quick to adjust to what works and what doesn't; willing to do more of something because others don't do enough, even if it's outside our comfort zone. It means being simultaneously pragmatic and principled, responsive, and adaptable. Luckily, as parents, we get lots of practice.[28]

There's also something else we can all do. We can communicate.

"Every DAY," says Katharine Hayhoe, leading climate scientist, "I do the most imp[ortant] thing anyone can [do to] change the system we live in: I TALK about climate change."[29]

Hayhoe has a huge platform. So do Al Gore, Leonardo di Caprio, and Alexandra Ocasio-Cortez. Sarah Myhre, who makes a career out of climate communication, knows where her own talents lie and has acted accordingly. "The communication stuff is part of my natural personality," she says. "I like people; I like talking to people. I have some good listening skills." I'm a philosopher and writer, so I'm writing this book, among other things.

But communication isn't limited to the big scale, and it isn't limited to "natural" writers or orators. This can be about tweets to millions of followers, huge speeches where you can change hundreds of thousands of minds in one go, but it can also involve chats about traffic pollution at the school gates or asking your physician if your child really needs the antibiotic they're about to prescribe.[30]

We all have networks: school and work communities, civil society organizations, like faith groups. We mostly have people who trust us, and whom we know well enough to know what matters to them. That's all important, because it's a lot easier to motivate someone if you start with something that they already care about. It's important because effective, authentic communication takes what Hayhoe calls "trusted members," such as family, friends, or members of the same church. (Hayhoe should know; she's an Evangelical Christian.) It's important because this shouldn't be about just one group of people.[31]

In practice, stances on climate change and COVID-19 have become a matter of political orientation. According to a recent study, voting Conservative rather than Liberal in the 2019 election in Canada was equivalent to tripling your carbon footprint overnight. In the United States, as in many other countries, both these crises have been turned into party hobby horses. But they really shouldn't be.[32]

I write this looking at a picture on my screen of a San Francisco pride parade. Among all the young people are a middle-aged man and woman, both smiling, rainbow banners round their necks. His sign reads, "I'm an old, white Republican and the Proud Father of a Gay Son." Hers says, "Christian Mom blessed with a wonderful gay son!!"[33]

This is what they get right. When our own children's futures are at stake, when human *lives* are at risk, we should all care. Right, left, or center. Young people know this already. Yes, the preferred policy details differ, but there is much more common ground than anyone would think, from listening to some older people, or their representatives. In 2018, more than half of millennial and young Republican or Republican-leaning adults thought the US government wasn't doing enough to protect the climate and environment. Seventy-eight percent agreed that the United States should prioritize renewables over fossil fuels.[34]

I reach out to Benji Backer, president of the American Conservation Coalition and self-proclaimed conservative. "I know young conservatives who are passionate about climate change and young liberals who are passionate about climate change," he tells me, "and others who don't have a political affiliation who also really care. We have a duty to protect the planet. We want to have life for the foreseeable future. We want to have kids and grandkids and give them a better world. We love nature and want to protect it. What's political about that?"[35]

While climate change is the most unanimous issue among young people, he says, young conservatives also care about social justice, race, and poverty. These are all reasons for their increasing disaffection with the Republican Party.

Politicians should take note.

BEING ALLIES, NOT "MOM SAVIORS"

In London in 2016, developers built a complex of luxury flats. In line with legal requirements, they also included some affordable housing. In the original plans, a gate let children from the social housing area into the development's garden and play area. When the development was finished, the gate had become a hedge. Only families who owned their own homes could play on the grass. The rest got a grim, narrow, artificial strip.

Children were baffled and upset, parted from their friends. Mothers from the homeowning and the affordable housing sections teamed up to demand change. The story was published in *The Guardian* in 2019 and quickly picked up by other outlets. Politicians condemned the developers. Ultimately, the gate was put in.[36]

The moral of this story? (Apart from some strong thoughts on the characters of developers?) Parents can do a lot. But here's something else for affluent, often white, parents in rich states to remember. If we really want to *be* allies, if we want to build a just world for our kids to inherit, we must work *with* marginalized or vulnerable groups, taking their lead. "You aren't allowed to make this about rich against poor," Louise Whiteley, one of the London mums, told a journalist. "This is just about a group of women, all friends, fighting together for children to play."

There are two dangers here, at opposite extremes. One danger is stepping back, seizing on some "hero," often from a marginalized group, heaping them with praise, and expecting them to save the world. Middle-aged, middle-class parents, changing nothing themselves but saying the school strikers "give them hope," do just that. The other danger is trying to be a "white savior," sweeping in and taking over.

This is about learning, changing, educating myself, accepting complicity. It is about being prepared to be uncomfortable, not defensive. "I cannot hide my anger to spare you guilt, nor hurt feelings, nor answering anger," says Audre Lorde, African American activist, mother, and poet, "for to do so insults and trivializes all our efforts." The role of white people at Black Lives

Matter protests, says Ben O'Keefe, activist and former assistant to Senator Elizabeth Warren, is not to loot or riot or exacerbate tension (for which Black protesters will be blamed, and further vilified). It is to "to do everything they can in their power to put their bodies between the bodies of black people and police."[37]

In 2014, a young woman posted on Twitter: "Guy at a hotel gave me 'that look' and said 'Helloooo' to me. His friend said: 'don't be a dick' – need more guys like that." We do indeed. We can also *be* more like that. We all have opportunities, every day, to call out destructive norms. Sociologists at the University of Tasmania tell would-be male allies: "Don't just dismiss sexist, ableist, racist, homophobic or transphobic comments or acts as 'banter' or jokes." Call them out.[38]

As a white woman and mother, I must listen as well as shout. I must use my social power (real, if not adequately reflected on) to insist that Indigenous mothers, mothers of color, mothers in the Global South, are seen and heard. They have been leading the way in demanding justice, for decades. And, as parents, we must hear and amplify the voice of the young people whose future is at stake, and who have already done so much to show us the way.

Chapter 5 in Essence

To build a safe, just future for our children, we need institutional, social, political, corporate, and norm change. This won't happen without social and political movements to bring it about. As parents, we have a powerful voice. Whatever our political affiliations, we are all humans, and we love our children. We must use that voice. Many parents already do.

What does this mean for me, or you, as an individual? It means thinking less about what I can do and more about what we can. Philosophically, we can make sense of this. By taking part, I might just trigger a huge change, I can help avert harm, and I can do my fair share in what we should all be working toward.

This is activism, but in a broad sense. It could involve anything from demanding that your bank or pension fund invest differently, to marching, campaigning, civil disobedience, or supporting social movements for plant-based eating. No one can do everything, so we should each weigh up our own skills, position, and resources, against these collective needs, and decide where best we fit in. We should all raise awareness, use our networks—and communicate.

6 RAISING GOOD GLOBAL CITIZENS

I want my children to be happy. I also want them to be kind.

It matters how we raise our kids. We know this. We encourage them not to hit each other, to say sorry and thank you, to think about how others feel. When I bid my girls goodbye at the school gates, we have a little routine. "Be kind," I tell them. "Do your best, have fun." I kiss them (if they let me). I ask: "Who's awesome?" Sometimes they smile, sometimes they're in a morning grump. The older one rolls her eyes. But eventually they say, "Me."[1]

I do this partly because I want to start our day on a good note—or rather, *restart* it that way, after breakfast tears, constant lateness, and quarrels over missing shoes—and partly for a more fundamental reason. This is the most important reminder I could give them: to be considerate to others and to themselves. But, in this challenged and complex world, those "others" can't just be each other, their dad and me, their grandparents and cousins, friends and teachers.

That's what this chapter is about.

I want my daughters to be good global citizens. I want to be one myself. As we've seen, that means being sensitive to others' needs, whether they are next-door neighbors or "distant strangers." It means working with others to build the institutions that will protect us all.

Turning out our children this way means two things. It means giving them facts, and it means giving them values. It means educating about them about climate change, antibiotic resistance, the reasons for the pandemic that shut down their world. It means telling them who is hurt worst by these things, and why. It means exposing them to what Al Gore calls "inconvenient" and the anti-racist movement "uncomfortable" truths. It means raising children who are kind and compassionate, who recognize *all* their fellow humans as morally significant, are motivated to demand collective change, and will play their part in bringing it about.[2]

So far, so vaguely grandiose. But *why* should parents, who are busy enough anyway, be doing this? And how on earth are we supposed to do it?

Let's start with the why.

BEING MORAL, BEING HAPPY

Serial killer Ted Bundy murdered more than thirty women. He relished his sadistic actions and never expressed guilt or remorse. Until he was caught, he lived the life he wanted, and he enjoyed it. But was his life a *good life*, at least up to his arrest? Was it even a *happy* one? Philosophers Jennifer Wilson Mulnix and M. J. Mulnix ask these questions. I'll put it another way.

Would you want your child to *be* Bundy, even a Bundy who remained uncaught and satisfied until the end of his days?[3]

The thought makes me sick.

There's a debate, almost as old as philosophy itself, about the relationship between individual happiness or flourishing, and moral goodness. Is Plato right, and "a good life" neither more nor less than a *morally* good life: a virtuous life, a life lived justly? Or do we need to follow a moral path for flourishing, but need other things as well? Aristotle thought so, and plenty of modern philosophers agree. Or are the two separate, so you can lead a life that's completely good *for you*, while doing terrible things?[4]

The truth, most plausibly, lies in the middle. It's counterintuitive to reduce human flourishing (let alone happiness) to moral conduct. We have needs of our own: capacities for pleasure, projects, and people we care about. We want to run on the sand barefoot, drink Sangria, play football, make scientific discoveries, spend time with those we love. As we saw in chapter 2, nobody wants to be a moral saint.

But could our own flourishing, the kind of flourishing we want to protect for our kids, have *nothing* to do with the moral claims of others? That's equally counterintuitive. As Mulnix and Mulnix put it: "If a monster like Bundy can be happy, then this seems to strip happiness of all interest and value for us."[5]

So far, you might think, so obvious. No one thinks we should raise our children to be Ted Bundy. But what I'm talking about is more controversial.

In 2014, the Pew Research Center surveyed American adults on the importance of teaching different qualities to children. "Responsibility" was crucial, with 93 percent saying it was "especially important" and 55 percent that it was the most important trait to cultivate in children. I agree that responsibility is crucial, but I interpret it broadly.[6]

I think I should teach my kids to respond to human suffering wherever it is. I think I should motivate them to work with others to help end that suffering, even if they are only part of causing it or could only be part of preventing it. I think we should raise feminists, anti-racists, and advocates for the global poor, social justice, climate justice, and health justice. My version of "responsibility" fits with "empathy for others," which 86 percent of consistent liberals and more than half (55 percent) of consistent conservatives ranked as important in raising children. But how does it fit with "obedience," which 67 percent of consistent conservatives ranked as important (and 15 percent of them as especially so)?

Remember Bob, who denied that his privileged kids needed a just world? Bob is no monster, and he doesn't want his sons to be Ted Bundy. That, he points out, is just a matter of raising them to obey the moral rules prevalent in their society. However, he objects to raising them to *challenge* social norms and accept responsibilities well beyond those recognized by most of the people they are growing up among. That, he says, will just make their lives harder.

Or put it like this. I said our kids deserve a world where they aren't continually torn between the moral pull of suffering others, and their own lives and interests. Now, I am saying we should raise our kids to identify more closely *with* their moral selves. Bob thinks otherwise. "I'm doing very nicely as things are, thanks," he says. "I do my bit for my family and friends, and I don't go around murdering anyone. I can give my sons the same advantages that I have. So why shouldn't I forget about changing the world, and keep them from inner turmoil by bringing them up *not* to think about these 'distant strangers' you're getting so worried about? That seems a whole lot simpler."

This argument is flawed. It's flawed *even if Bob only cares about his own sons*. As we've seen, there's plenty of empirical evidence that a fundamentally just society, not a divided one, is better for everyone. Climate change, antibiotic resistance, and global pandemics are dangerous across the board. They will be bad for our children—mine and Bob's—if they aren't already. They will be worse still for theirs. Closing our eyes to global suffering has proved to be a disaster all round.

I could say all this to Bob. I could also point out that he is, most likely, *morally* inconsistent.

I could ask him: "Suppose your son could be Bundy but live in a world of Bundys and victims: a free-for-all where the strong torture, unpunished and delighted, and the vulnerable are tortured. Would you really want that?" If Bob says no (which he must, or he *would* be a monster), his sons' immediate satisfaction is not all he cares about. Once he's accepted that others matter, that he and his boys have the two moral duties I began this book with: not to do serious harm to others, and to help them if they easily can, he must face up to what that implies in this interconnected world. Even if he only accepts that they shouldn't do major harm to others, the ramifications are vast.

In other words, Bob's children, like mine, will grow up to be global citizens, like me and—whether *he* likes it or not—like Bob.

This is important in an obvious way. Our kids will play a key part in determining the future for others. This is especially true if they are comparatively privileged, as mine are. As global citizens ourselves, we cannot forget that, when we are raising them. Even if Bob's preferred option (unjust future, unbothered kids) would spare them psychological marring, it would still be morally wrong.

It also matters in a more subtle way. Here's something else we've already seen. The parent-child relationship is fundamental to our children's prospects in life. So suppose I do what I should (and Bob should!), as a moral agent. I become an activist, do my best to build a better world. But I don't involve my girls, explain what I'm doing, or encourage them to do the same. How will they feel when they are old enough to understand?

I think they'd feel betrayed. This isn't the same as failing to share other interests I might have: writing books, cycling, or singing along badly to 1980s pop. In fulfilling these moral duties, I recognize myself as part of a global community, a moral person among other moral people. If I don't raise my girls to do the same, I'm not acknowledging them as members of the same community. I'm not recognizing them as truly human beings.[7]

The same applies to Bob. Even if he avoids this disconnect by *not* fulfilling his basic obligations to others, this double failure is a betrayal. His sons may feel the pull of the basic needs of others, despite his best efforts. Then they'll resent their dad for not helping them respond, for making it harder for them to do what's right and putting them on the wrong side of history.

Jennifer Harvey, anti-racism educator, describes a common reaction among nineteen-year-old white college students. They are angry because, in being taught to be color-blind, they have been left unprepared for the reality of racial injustice. They are left flailing, unable to relate to their fellow students of color. "Why wouldn't you tell me this sooner?" they ask.[8]

"*Why*," Bob's sons could demand, if he raised them as members of an unthinking elite in a desperate world, "wouldn't you tell me this at all?"

EMPOWERING, NOT SCARING

There are two major concerns about raising our children in this way. The first is this: When does moral education end, and brainwashing begin?

Aged four, my daughter demanded that her father draw her a picture. What of, he asked. She named a senior Tory politician, and said she wanted him being eaten by a crocodile. I laughed when my husband told me, but it also prompted a mother-daughter talk and a mental note to tone down my own political rhetoric. My editor is similarly uneasy when she describes her (much older) children parroting jokes about Donald Trump. I want to raise morally aware, globally conscious children who will call out injustice. I don't want to increase hatred or polarization.

And I don't want to stop them thinking for themselves.

We humans are autonomous beings. We flourish, in part, through living our lives for ourselves. To do this, we need both the capacity for rational reflection and deliberation, and the external circumstances to make it meaningful. In other words, we need an adequate range of options. Our parental authority, remember, isn't ours by unassailable right. It comes on the condition that we use it to fulfill our moral responsibilities to our kids. And that, among other things, means cultivating their autonomy.[9]

Because of this, some philosophers worry about children being taught so-called comprehensive views, such as stringent religious doctrines, because that could restrict their future freedom of choice. In raising my girls to be passionate about climate and social justice, to demand action on global emergencies, am I limiting their autonomy? Am I trying to make them mini-mes (or, rather, mini versions of what I think I *should* be) and depriving them of the chance to be themselves?[10]

The second concern comes from another direction: I don't want to scare them.

My daughters play. They run and climb and do headstands. They cry, they fight, they laugh. They paper the world with their imaginations and build beautiful, towering castles in the clouds. They are children.

As philosopher Samantha Brennan puts it, being a child is not just a "prep school for adult life." It is easy to forget this, in rushing to prepare our kids for that adult future. But Brennan tests us. She asks whether we would be happy for our child to be given a pill to turn them instantly into an adult, with all the education and development stages neatly ticked off. I wouldn't, and it's not just because my girls' childhood is a wonderful, too-short period *for me*. I feel instinctively, deeply, that I would be doing them a great wrong.[11]

In his autobiography, the poet John Betjeman describes wild weather, waves splashing the bungalow walls, and an all-pervading sense of security, snug inside. I read those lines for the first time comparatively recently and felt immediate recognition. The feeling he describes is the one I had as a child, (ironically enough) in the back of a car. Wrapped in my sleeping-bag cocoon, entirely trusting my mum or dad at the wheel, I relished the slamming rain, the dark outside. I never thought of the dangers we faced, or how hard it must have been for them.[12]

Some things are valuable in any period of life, but it matters particularly that we have them as kids. As I observed in chapter 1, you cannot make up for a childhood without love or affection, or space to run, by having an abundance of these later in life. Other things can *only* properly be enjoyed in childhood. Philosophers call these "intrinsic childhood goods." As children, says Brennan, we can trust someone else absolutely. Children, truly to *be* children, need to experience that trust along with a heady cocktail of carefreeness and potential.

Poets recognize this. Not just Betjeman, but also Helen Dunmore in "To My Nine-Year-Old Self," and Emily Dickinson in "The Child's Faith Is New." Philosophers recognize it too. Everything seems possible in childhood, says Brennan. Time appears endless. There is a sense, giddy in its limitlessness, lost in adulthood, of having "one's whole life stretched out ahead." "Innocence," says Colin MacLeod, "permits various childhood choices and new discoveries to be accompanied by a sense of wonderment and joy." Such innocence, he thinks, *requires* some ignorance of how the world is.[13]

As my daughters have grown, parenting has become complex. My instinct to shield them goes unsatisfied, and I must teach them to deal with what the world throws at them: wounds that cannot be solved with a kiss from Mummy or a fistful of chocolate buttons, problems at school, friendships to negotiate.

As they become teenagers, then young adults, these obstacles will only grow, and my ability to resolve them diminish. I know that already, and it terrifies me.

It's a dilemma. As we've just seen, part of my job is to help my children grow out of their carefree state and develop the adult agency they need for an adult world. But if they lose too early what Dickinson calls their "pretty estimates / Of Prickly Things," they will have lost something they can never get back. The last thing I want to do is to make them afraid.

The worry about brainwashing can be quickly dealt with. Frankly, I'm not too concerned about the damage to my daughters' autonomy, or ability to lead their own lives, if I turn them out strongly disinclined to work in the fossil fuel industry or support white supremacy.

Autonomy doesn't require infinite choice. If it did, no one would be autonomous. In raising them as good global citizens, I narrow my girls' options, but I still leave them with plenty. They can train as writers, dancers, doctors, or electricians. They can seek spiritual sustenance in church or in canyons. They can pursue political or social change *as* their career, or on top of it as activists, politicians, writers, voters, donators of money. All I'm pushing them to do is this: not choose a path in life that actively contributes to harm and fails to combat it.

It's also not at all clear that this valuable capacity, autonomy, must include maximum freedom to make morally terrible choices, or to pursue injustice. (Is Bundy's autonomy valuable?) Even if it must, autonomy is just one part of human flourishing, not all of it. "While parents have a duty to respect their child's future autonomy," says the moral philosopher Tim

Fowler, "they also have a duty to protect their children from living badly." And I'm *not* recommending that parents pass on a deeply restrictive world view: just that we teach our kids not to "live badly" in the most fundamental way there is.[14]

That doesn't mean parents shouldn't share other causes and campaigns important to them with their kids, so long as those aren't actively opposed to justice or to treating their fellow humans as humans. As it happens, I think there's a balance to be struck between building a relationship with your child by doing things you value with them and maximizing their future autonomy. But even if I didn't, sharing the drive to fight basic injustice is a special case, because we should all be doing it![15]

One political theorist, Matthew Clayton, thinks parents shouldn't sign kids up to any "comprehensive" moral view, including religion. But he still thinks they should be taught the core values necessary to live together in a just society. In raising our children as good global citizens, we're giving them the core values they need to live together at all, as moral beings. In taking my kids on climate or anti-racism protests, I help them become *more* human.[16]

That leaves the other concern: scaring our children. It's deeper-rooted, but there is a blunt enough reply. The real crisis is what the world *will* throw at our young people, not being told (age-appropriately) about it. Perhaps some decisions should be made for kids—for example, in politics—because we should not "fast-track" them to adulthood. But the political decisions that now are being made for them are *not* in their own interests.[17]

Perhaps, then, we need "anticipatory parenting": warning our kids about the risks they face, helping reduce those risks, and teaching them to negotiate or withstand them. Being carefree can be liberating, says philosopher Sarah Hannan. But it is perilous if it means never properly appreciating what your actions might lead to. Preserving innocence isn't such a good thing, if it means keeping our kids ignorant of danger in a society that will not always be kind to them, even when they *are* children.[18]

In fact, *if* we care about autonomy, we have to respect that capacity as our children develop it. Kids have voices. They have growing agency, as well as interests, and they have a right to have their voices heard, especially as they get older. "Children need to be empowered," says Ugandan writer-activist Herbert Murungi. "They need to relate to the disasters happening in their own environment."[19]

Besides, what's the alternative?

THE NEUTRALITY MYTH

When my daughters were babies, I dressed them in hand-me-downs from my nephews, much to the consternation of older ladies on buses. One exclaimed in confusion at a pink hat and blue coat combination, unable to guess the baby's sex. "But she's wearing blue!" another one protested, when I said my child was a girl. She could hardly have sounded more accusing if I'd been in the dock for fraud.

The incidents were minor enough. But the point they illustrate isn't. There's no such thing as neutrality, in the world our children are growing up in. They will acquire values from somewhere, and if they don't get them from us, they'll pick them up elsewhere.

Children, says Harvey, inhale racism "like smog in the air." According to behavioral scientist Pragya Agarwal, they have well-formed attitudes by about six. By nine, highly stereotyped views can be hard to root out. In a racist society (and our societies *are* still fundamentally racist), not teaching them to be anti-racist is, in effect, leaving them to become the reverse. If we don't raise our kids to be feminists, they will learn their own lessons from the myriad incidents of sexism or misogyny that they see around them.[20]

Kids are also exposed early to consumerism. With almost every interaction, they learn to value a collectively disastrous way of life. I didn't want my daughters to be materialistic; I certainly didn't set out to make them so. But it didn't take them long to demand shiny plastic toys in the likeness of their favorite cartoon characters, or the same gadgets as their friends. Consumer psychologist Cathrine Jansson-Boyd tells me that teenagers learn quickly to label people, in terms of what they can and cannot afford. "Parents," says Fowler, "need to decide what values children will be socialized into endorsing."[21]

There's something else I should remember, too. The secure childhood I enjoyed, and yearn to reproduce for my own kids, was also a sign of my own privilege.

"Black children don't get to not understand racism." That's what comedian Tawny Newsome tweeted, nine days after George Floyd's murder. In 2014, twelve-year-old Tamir Rice was shot by Cleveland police because he was playing with a replica gun. Writer Sai'iyda Shabazz won't even let her six-year-old son play with a water pistol in public. In 2020, a Young-Minds blogger, Luke, aged fifteen, wrote: "I felt so depressed and almost worthless in myself because of my skin color." He was the victim of racist bullying.[22]

"It's deeply necessary," says Jennifer Harvey, "that we let our children's hearts get broken a bit if they are going to remain able to recognize the humanity of their fellow humans whose lives are at stake in the system we live in." White parents, she says, teach white kids that police officers are safe. They tell to go to them if they are lost. Black parents can't do that. When political theorist Alice Baderin coined the term "anticipatory parenting," it was to describe what Black parents must already do, day by day: trying to spare their child racist experiences (for example, by constant vigilance or making sure they aren't the only child of color in a group); trying to mitigate the mental and physical harm of the incidents that are, despite this, inevitable.[23]

Harvey tells another anecdote: her daughter being told by a boy at school that he could cut ahead of her in line because "boys go first." Her child, she says, "was having an experience of sexism whether I decided to support her in interpreting it or not." Our daughters will learn soon enough that this is a man's world, even if we don't tell them. They will learn it from incidents like those. From reading "Boys" (in blue) on the boxes of engineering toys and "Girls" (in pink) on the ones with florists or hairdressers. From clothes that sexualize twelve-year-old girls or turn teenage boys into pretend soldiers. From seeing powerful male politicians get away with harassment. They will learn to expect groping on public transport, or sexual harassment in the playground. "If I had a penny for every sexist thing I've heard kids say," says Melinda Wenner Moyer, in her brilliantly titled book *How to Raise Kids Who Aren't Assholes*, "I would be lounging on a beach in Bermuda rather than writing this book."[24]

In 2019, UK psychologists warned: "For children growing up in a landscape of ever-increasing danger and parental stress, we risk developmental trauma becoming a 'normal' part of childhood experience." In 2019, 57 percent of teenagers in the United States were afraid of climate change. More than half were angry. And that was before the pandemic. In 2020, a UK children's news program interviewed two thousand eight- to sixteen-year-olds. Almost three quarters were worried about the state of the planet and 22 percent were "very worried." Seventeen percent struggled with eating or sleeping because of it, and four out of ten didn't trust adults to tackle the problem. In 2022, a seventeen-year-old Australian activist, Anjali Sharma, published an article in a major newspaper: "Dear politicians, young climate activists are not abuse victims, we are children who read news."[25]

In the face of this, silence isn't "being kind." It's dangerous. Susan Clayton, psychology professor and fellow of the American Psychological Association, puts it starkly enough: "If you try not to talk to your child, they will still know it's a problem. They will just think this isn't something parents can help with. It will erode their feelings of confidence in their parents."[26]

In my snug bubble, a child in the back of my parents' car, the last thing I was worrying about was what all the cars on the road were doing to the planet. I didn't know about the communities in Nigeria, choking on pollution, lands ransacked by Shell, to provide the fuel. But I know now.

So, yes, it breaks *my* heart, to talk to my girls of floods and wildfires, Russian military tanks and families fleeing their homes, gun deaths and hospitals that cannot keep their patients safe. I hate to see fear in the faces that I want to see laughing. And, yes, this is an additional wrong that we're doing our kids, requiring them to respond to the mess that past generations have created. But pretending that it isn't happening? That's a betrayal too. Harvey is right: this is needed, to prepare our kids for the world they are growing up in.

Like most of parenting, this isn't an all-or-nothing choice: tell the whole truth now or keep our children's childhoods idyllic. There's a fine line to tread between preparing our kids for the future they will actually face and not piling more harsh reality onto them than they can cope with; between building them up into good global citizens and keeping the joy of those early years, as far as we can.

So how do we go about it?

WALKING THE LINE

Point one. Being good parents means having hard conversations—and making our children part of the way forward.

After years of happily wearing whatever color came to hand, my older daughter returned from school one day and told me blue was a "boy's color." I told her there was no such thing. I reminded her of the boys we know who love pink, of how often *I* wear blue. Researching for this book, a few years later, I realized I should have done something else. I should have warned her how often she would meet that kind of prejudice. When Harvey's daughter first talked about the boy pushing ahead at school, Harvey avoided using the language of "sexism." But later, with her daughter still struggling to process what had happened, she gave it a label. She warned her to expect many such incidences, and told her that in standing up for herself, she had stood up for other girls.

White parents, says Harvey, should have a version of "the talk," preparing their children for what their contemporaries of color will face, and building them up to be allies. Remember the teenagers she worked with, angry and perplexed when they learned the truth? Color-blind parenting doesn't work when the world around your child is divided by color.

But in presenting these unpalatable truths, it's essential to distinguish between individual and collective responsibility. Harvey's daughter, confused by another incident at school, asked her, "Am I racist?" But it's the system we live in that's racist. If your children are privileged—and mine are, I know—they must learn to understand that, without hating themselves for it.[27]

Early in the Russia-Ukraine war, Pragya Agarwal shared her advice on talking to kids about the situation. "Be honest, explain and don't brush away their questions, stay calm, try not to show fear & anxiety, listen actively *and think about positive things we can do together*" (italics added).[28] Because if our children *are* scared (and they often are), then getting involved, from raising money for refugees to marching outside government buildings, is one way for them to regain a sense of control. In encouraging this, says Susan Clayton, you simultaneously do something positive, and boost your children's mental health.

When my older girl was five, she announced that she and her friend were going to make a machine to take all the plastic out of the ocean when they grew up. She was excited. It wasn't a burden, at that age. It was an opportunity. For months after their first climate marches, both my girls would burst into a spontaneous chant: "What do we want? CLIMATE JUSTICE! When do we want it? *NOW*." They did it at disconcerting moments: on a bus, in the post office, halfway down a busy street. But I was glad they did.

Point two. To quote educator Harriet Shugarman, we must "make time to read, share stories, and be together in nature."[29]

Herbert Murungi combines all three. He tells me how his parents sat round the fireplace with him, telling fairy tales. Now his own children's stories are distributed to schools, and he fundraises to plant trees with the kids that read them. In Cape Town, Xoli Fuyani "didn't want to go in and lecture, we wanted to create something fun and experimental." She ended up with worm farms, and the school kids loved them. At home, she grows most of her own vegetables, and has her own children learn by helping out. In Vermont, Meredith Niles, an associate professor in food systems, describes her child's joy in their garden: how the little girl will pick a tomato and bite into it as though it were an apple. Studies confirm what these educators and activists all know: access to nature encourages kids to behave in an environmentally friendly way.[30]

Equally, it takes more than a generic "Girls *can* do that!" to counter gender stereotypes. "Look at your aunt," we must say, pointing to a doctor or engineer or scientist. "Look at your friend's mother or your physician." Our children need examples of activists of color, says Harvey, *and* of white people fighting for anti-racism. They need true stories. They need fiction. They need stories featuring children of color that have nothing to do with racism.

I try. I read my girls *Goodnight Stories for Rebel Girls*. I read them the *Unicorn Rescue Society* and *Zoey and Sassafras*. When we read the books I loved unquestioning as a child, I help them call out stereotypes. It's a start.[31]

Point three. How we do all this matters, too.

In the novel *Hard Times*, schoolmaster Thomas Gradgrind raises his children like he teaches those of others: according to a strict formula. The beauty of flowers counts for nothing, hard statistics are all. He condemns emotion and shows none in his own home. The result, for his children, is disaster. His son becomes a thief; his daughter marries a bully she doesn't love, contemplates infidelity, and is thoroughly miserable.[32]

The moral of Charles Dickens's story, now neatly confirmed by developmental psychology, is this. *Authoritative* parenting—proactive, supportive, sympathetic, using positive reinforcement and honest dialogue, setting limits and clear rules, but not applying them arbitrarily—is better for moral development than a harsh, punitive, or power-assertive *authoritarian* approach.[33]

"Our children are watching us," Shugarman tells me, "even if they don't seem to be listening." Psychologists Mary Eberly Lewis and Scott Franz studied conversations about volunteering between fifty-two teenage girls and their mothers. They found the teenagers were less receptive if the conversations were disruptive or antagonistic, if their mothers dominated or tried to control the debate, showed contempt, or devalued the project. " A 2015 study found that parents with higher "cognitive empathy" (the ability to take in others' perspectives) had toddlers who were better at sharing.[34]

As a parent, I can model speaking up. I can model care for others. I can model care for nature. I can help our children learn to take disappointment in stride, and to feel pride in action. I can encourage my daughters to ask questions, ask their opinions, and help them work through difficult ideas for themselves. I can talk climate action over the dinner table, as Shugarman suggests, and really listen to what they have to say. When I take them on climate or anti-racism marches, we can discuss why it matters so much, along the way. These interactions are Socratic, rather than laying down the law. They also forward my daughters' growing autonomy.

I can be there for them. "Not feeling alone," says Sarah Jacquette Ray, "is probably the most important prescription for long-term resilience." If our children are to cope with what the world will (and we must) teach them, they will need to learn how to manage, not deny, the emotions that these ongoing crises throw at them. This will not be easy. They need love and compassion, empathy, and availability. They need to learn to develop friendships. My daughters need to know, now and in the future, that they have "adult allies" and that I am one of them.[35]

The key to all this? Different approaches, for different ages.[36]

In the early years, climate education is about enabling kids to experience the wonder of nature, about love and hugs ("if," as Shugarman puts it, "that's your thing"). Kids might also be taught to respect shared resources, whether it's things at home that belong to the whole family, or equipment at the playground in the park. Encouraging empathy means helping preschoolers express emotions, and reflect on those of others. It means asking, "How do you think that made her feel?" Anti-racist and anti-sexist education begins with openness about difference, teaching children that they are in control of their own bodies, modeling consent.

Then there are the elementary school years, when we can point out examples in books or TV that go against stereotypes (a cartoon Dad staying at home with his kids, a Black woman scientist), or supplement the often-limited version of history taught in schools. We can label feelings, challenge disrespectful language, and encourage our kids to do the same. We can take them with us on protests or give them simple jobs, like putting bottles in the recycling, or turning off lights. "So," says Clayton, "they feel engaged in the struggle."

As kids grow, we can share simplified science. We can talk about treasured outdoor places, why they matter, and how they are threatened. We can give names to key concepts like "consent" or "assault." We can encourage our kids to be friends, non-romantically, across genders. We can talk justice over dinner, be open about difficult topics like identity and oppression, prompt our children to notice power dynamics. We can encourage them to do school projects on justice or the environment.

As they get older again, kids can learn more complicated science and politics. They can think critically. They can be taught to recognize credible, peer-reviewed research, and critically evaluate their news sources. We can remind them that they, like us, are moral agents. "Help your high schooler understand that the world does not revolve around him," writes clinical psychologist Bobbi Wegner, on raising feminist boys. We can support our teenagers to become allies and activists, advocates rather than bystanders, and to make regular activities of challenging norms.[37]

The conversations with our children are very different, from ages three to eighteen. But the imperative is the same.

WHY PARENTS?

Fresh from my second maternity leave, I asked a colleague to read a paper I'd written on why we should raise environmentally conscious kids. Afterward, he told me, "You can tell your children are still young." He was right. My babies were three and one. I was the center of their world as they were (and are) of mine. Crafting my arguments, I'd assumed we could mold our kids as we please. The only question was whether we *should*. My colleague had teenagers; he knew better.

Myriad others can influence our children's capacity for good or bad: teachers, activity leaders, other family members, and family friends. They are swayed by the media, social media, and their peers, especially as they get older. But we *do* matter— and those others *aren't* doing what they should.

"Parents," says the adolescent psychologist Daniel Lapsley, "are clearly the primary and most important source of moral intuition." Developmental psychology shows that we influence our kids' moral reasoning, their moral values, how well they see different perspectives or develop moral emotions like sympathy, and whether they act to benefit others. There's less empirical evidence on moral identity, but the theory suggests that supportive families make a difference there too.[38]

Psychologists Rosemary Randall and Paul Hoggett interviewed ten climate activists for a 2019 paper. Activists, they found, often connect social responsibility to family or cultural background. Psychotherapist Robert Tollemache found that Londoners with environmentally minded mothers were predisposed to pro-environmental thoughts and feelings. One man, who saw himself as largely detached from environmental concerns, described his father as "rather contemptuous" of early green arguments.[39]

What's more, if parents don't take on this task, and work together to get schools and politicians, corporations, and media giants to change what *they* do, it may not happen at all. "The history that British schoolchildren are taught is a sugar-coated, whitewashed version which focuses on the 'good bits'," warns Agarwal: "wars fought and won, and the white men who were victorious. . . . A quick review of the school curriculum shows a distinct lack of information about British colonialism, the imperial rule and its impact on the colonies, and Britain's role in slave trade." As I write, nine US states have passed legislation that would, in effect, ban teaching school children about racism. A majority of Americans, across all states, think schools should teach global warming, but legislators in several states continue to debate the question, influenced by climate denial organizations.[40]

My girls became aware of climate change because they were marching with me for climate justice. That gives me hope. I held them, tears in my eyes, while Kamala Harris stood by Joe Biden on a podium and said: "While I may be the first woman in this office, I will not be the last, because every little girl watching tonight sees that this is a country of possibilities."[41]

A world of possibilities. But only if we, *and they*, can make it so.

Chapter 6 in Essence

As parents, we owe it to our kids (and to everyone else) to raise them as moral agents. If we don't, we fail to acknowledge them as full human beings. In the world as it is, that means raising them as good global, intergenerational, and ecological citizens. We're not the only ones who should be doing this, but when schools and governments fail, parents must pick up the slack.

Will we brainwash our children by doing this, undermining their future autonomy? Not if we do it carefully. Will we scare them? We might. But neutrality is a myth. The kids at the receiving end of these ongoing injustices don't get not to know about them. If we don't raise our kids to challenge the status quo, then society will teach them to uphold it. Besides, they are already scared. If we don't acknowledge that, they'll only stop trusting us.

To walk the line between preserving our kids' childhoods, and empowering them to build a better future, we need hard conversations, stories, and time outdoors. We need to point out role models and be them ourselves. We must share ideas and listen to theirs; we must recognize them as the growing individuals they are. We need to be there for our kids, loving and authoritative, and never unreflectingly authoritarian.

7 LIVING DIFFERENTLY

We do the school run by electric cargo bike. The girls scramble on automatically now, the eight-year-old vaulting up as though she belongs in the Cirque du Soleil, the little one wriggling like a giant glow worm through the protective bars. We slide past traffic jams, kids chattering. We shrug off helmets and reflective ponchos at the school gates. Sometimes, I take my younger daughter and her friend, and they make a fairground ride of it, shrieking with joy at every pothole.

Occasionally, it's a pain. Cars passing too close. A taxi blocking the bike lane, so I'm forced into a hill start in the wrong gear. Cycle lanes that end abruptly, delivering us into the stream of traffic. The kids having a fight, wobbling the bike at traffic lights. Or the Scottish weather doing its thing, so we're battling horizontal rain, head to toe in waterproofs, lips wind-chapped, face and shoes drenched. But we still do it unless there's ice on the ground. This chapter is about why. It's about why (and how, and *which*) individual lifestyle changes matter.

But first, here's what I *don't* want the takeaway to be: "green" your own life and your children's, buy fair-trade clothes and chocolate, FSC certified toys, and don't worry about anything else. That would contradict everything this book has already said. Lifestyle changes—the right ones, the big ones—matter. But that doesn't make them the priority.

Reason one. In worrying only about our own lifestyles, we buy into something that governments and corporate giants would love to pretend is true, but absolutely isn't: that these global emergencies are a problem for individuals to solve.

Want a master class in this? Look at coal, oil, and gas giants. Even the US average individual carbon footprint, 17.58 tonnes annually, is a drop in the ocean compared with the 493 billion tonnes of carbon dioxide and methane that the top twenty fossil fuel corporations churned out between 1965 and 2018. Yet, when social scientists examined climate change documents from Exxon Mobil, they found a careful and disturbing use of rhetoric, designed to "individualize" responsibility.[1]

In fact, achieving change through individual lifestyle changes is harder, less fair, and less efficient than doing it collectively (and it's sometimes straightforward impossible). Even for those who can afford it (and many can't), it's difficult to ditch the car when public transport costs a fortune and cycling is unsafe. It's hard to go vegan when supermarkets sell beef burgers for half the price of vegan ones, or to avoid plastic packaging when almost everything comes in layers of it.

According to two recent reports, giving money to well-chosen emissions-cutting charities or investing it in "green" pensions or investments, cuts carbon many times more than almost any individual lifestyle changes. Some individual efforts might even be "canceled out." As Mike Berners-Lee points out, so long as there's only a finite amount of renewable energy, my family buying *only* renewables could just mean another family, with a different tariff, using less of them.[2]

This extends, perhaps especially, to that most emotive and controversial of "low-impact" life choices. In focusing political or social attention on whether individuals and couples should have kids, and how many of them, it's easy to forget *why* having a child in a country like ours adds to environmental devastation: the lack of effective net-zero policies. Compare the expected carbon increase from having a child in France (1.4 tonnes per year) to having one in the United States (7 tonnes). Then think about how much more that figure could be reduced, by the right policies.[3]

Reason two. This burden falls especially on women, worsening gender divides. That's most obviously true when the conversation turns to having babies, which in effect means passing judgment on how women do (or don't) use their own bodies. It's also more generally true, since so much of the work of "greening" households tends to fall on women, who still do the lion's share of home organization, shopping, and cooking. (Having said that, according to Maya Mailer, most parent climate activists are also women, so calling for activism isn't unproblematic either, unless it comes with a serious effect to redistribute domestic burdens.)[4]

Promoting collective change takes priority. So where, and how, do individual life changes fit in? And *what* changes are we talking about?

Let's start with the second question.

RESPECTFUL INTERACTIONS

Being a moral person—and bringing your kid up as one—is about recognizing and respecting your fellow human beings *as human beings*. Whatever level of citizenship we're talking about, that's a prerequisite. This applies to individual, day-to-day interactions, too, with implications that those like me, sheltered by privilege, may not have recognized.

Being *anti*-racist or *anti*-sexist, and raising our kids to be the same, means doing as well as not doing. It means "not being a dick," like the man called out by his friend in chapter 5. It means avoiding what some call "microaggressions" but professor and anti-racism activist Ibram X. Kendi calls "racial abuse." Obviously, it means *not* phoning the police just because a Black man is watching birds in Central Park. It means not always sitting next to a white person on the bus. But it's also more complicated than that.[5]

It means actively questioning our own unthought behaviors, unlearning some habits and patterns and making darn sure we don't pass them down to our kids. "See your situation clearly," say University of Tasmania sociologists, advising men on how to be anti-sexist. Educate yourself on your own privilege and biases. Read books, articles, and blogs (there are plenty out there). Go to diversity and inclusion events. "Do not rely on people in marginalized groups to educate you."[6]

Showing respect means anticipating how certain behaviors could be perceived as a threat, by those who are used to being threatened. This can be as simple as a man crossing the street to avoid following a woman on a dark night. It may be a pleasant jog for him, but for her having an unknown man behind her means walking faster and faster, knuckles tight on keys, not wanting to turn around.

Showing respect also means *not* indulging in our own discomfort if that discomfort is born of privilege. As the Tasmanian sociologists put it, "It's not about you." Activist Ben O'Keefe got frustrated by white friends making a gesture of asking how he was doing in the wake of the George Floyd killings. Or, still worse, making it about how scared and ashamed and sad *they* were. "Part of being an ally," he says, "is taking a deep breath and getting past the shame and the guilt that you're carrying." Want to reach out to Black friends after another racialized attack? O'Keefe suggests a simple: "Hey, I can't imagine what you're going through, I'm here if you need it." And mean it.[7]

MODERATION, NOT MATERIALISM

Respect is—or should be—absolute. But what about the other things we do and use, day by day?

Recycle, reuse, turn down the heating, buy energy efficiency technology, buy green energy, conserve water, maybe drive an electric car—and boom, you've tackled climate change. That's what governments would have us believe. Or so scientists Kimberley Nicholas and Seth Wynes discovered when they surveyed Canadian high school textbooks and government resources from the United States, Canada, and the European Union. Politicians not only like to make this all about individuals, they also like to advocate round-the-edges tweaks that won't ruffle feathers among vested interests.[8]

That's unfortunate because most of these aren't the big-ticket lifestyle choices, as far as carbon emissions are concerned.

There are four of *those*, according to Wynes and Nicholas: don't fly, go car free, go vegan, and have one fewer child than you would otherwise have done. The last one is complicated—and we've discussed it already—so I'll focus on the other three.

"Travel fresh" and "holiday local," says grassroots movement Take the Jump, based on research by the University of Leeds on achieving sustainable urban living. Ditch the private car for cycling, walking, public transport, and ride-sharing, and take just one flight every three years. According to Founders Pledge, a global community of entrepreneurs who commit to give a portion of their personal proceeds to charity, shifting to an electric car saves two tonnes of carbon a year, while giving up your car altogether saves 2.4. Taking one return transatlantic flight comes at a carbon price of 1.6 tonnes. For an internal European flight, it's 0.08 tonnes.[9]

When it comes to diet, it's all about the plants, but not about *all* plants.

Not everyone advocates going completely vegan, but there's a clear consensus that most people should eat more plants and a lot fewer animal products. The top carbon priority, according to Mike Berners-Lee, is cutting down on cow and sheep products, while you can help combat antibiotic resistance by not buying from farms that give preventative antibiotics to healthy animals. (If you're wondering how you can tell, Berners-Lee is pretty blunt about that too: "It is fairly safe to assume the worst unless you know otherwise."[10])

But did you know that rice from flooded fields has twice the carbon footprint of the equivalent calories in oatmeal or potatoes or bread, and three times as much as maize? I didn't, until I saw a study of nearly forty thousand farms, confirming it. And as smug

I might like to feel as a vegan, I can't. Not when dark chocolate has an average carbon cost of 2.3 kilograms per 50 grams.[11]

How our food gets to us matters too, though it's not always as simple as "local good, everything else bad." "*There is no place for air freighted food in the twenty-first century*," says Berners-Lee. He recommends checking the country of origin, then figuring out whether the food would last long enough to travel by ship, train, or lorry. Bananas, apples, oranges, yes; strawberries, grapes, asparagus, no. But food hot-housed locally can be as bad as flying. Scottish strawberries in January? Almost certainly another no.[12]

About a third of food is wasted and the majority of that, in high-income countries, is in restaurants and households. "Individuals have to be part of the solution," says Meredith Niles, food security expert, speaking to me from Vermont, "composting or, even better, avoiding food waste in first place." And, perhaps, watching portion size. The average human, says Berners-Lee, consumes 2,530 calories daily, 180 more than they need for a healthy diet.[13]

Then there is the small matter of breaking the consumerist habit of several lifetimes.

Here's what I can do. Buy better. Choose ethical clothes, FSC wood, Fairtrade food. Buy less *new stuff*. "Dress Retro," says the Take the Jump website. "Three new items of clothing a year." We can use thrift shops, eBay, local secondhand marketplaces. We can hang on to products for at least seven years, as Take the Jump recommends, rather than constantly upgrading.

Most importantly, we can have less stuff altogether, and make sure our kids do. "Be an assertive parent," advises consumer psychologist, Cathrine Jansson-Boyd. "Say no and be firm with the no, and explain why." It seems an impossible aim, in the face of constant, peer- and TV-driven demand. It's incredibly hard to say no to my kids and, frankly, I'm not very good at it. But Jansson-Boyd assures me that it will start to work, over time.

And, of course, if we are lucky, we can give our kids activities in place of these endless *things*. Time playing with them. Time spent outdoors; giving our children the smell of the sea and the sight of the diving birds; the feel of mud and stones under their bare toes.

So much for *what* we must do. Now for the *why*.

FAMILY LIFE AND COLLECTIVE CHANGE

"Activism can come with a feeling of having to fight and grate against the system," says Sarah Myhre, climate communicator, "and I think that is well and good, but the deeper truth is that you have to care about people, about places. That's about changing who we are as people."

This might seem straightforward, and on some philosophical accounts it is. Virtues like temperance or treading lightly on the earth involve exactly kind of changes I've just described. Integrity, for some philosophers, means reflecting on your own behavior, the changes you think society must make. But, for others, there remains the same instinctive, rational barrier that we encountered in chapter 5, with regard to activism. Why overhaul the family lifestyle when it seems to make no difference?[14]

Luckily, we can use the same tools here, to explain why. For one thing, there's a small chance of making a big difference.

Philosopher Shelly Kagan offers an example. Suppose you want to buy a chicken for dinner. It might not make a difference. But what if the store needed a certain number of chickens to be sold in a day before it would order more, at which point the supplier would up their production? Suppose that happens when 25 chickens are sold in a day. So, if 25 rather than 24 go off the shelves on any given day, the supplier kills 25 more chickens, and another 25 eggs are incubated so they can be hatched and raised in hideously painful conditions, to produce another 25 chicken dinners.[15]

It's an oversimplification, but the general point holds. You know, Kagan says, that there's a 24 in 25 chance that your purchase won't make any difference to the number of chickens living horribly and dying painfully. But there's a 1 in 25 chance that your decision will condemn 25 to this, so the "expected chicken deaths" associated with it comes out at 1 in 25, multiplied by 25. One chicken suffers and dies for you.

I haven't asked you to assume that the suffering and death of a chicken is bad in itself, let alone bad enough to outweigh the pleasure you get from eating it (although, as it happens, I think that both are true). But I *have* assumed that human suffering matters. Men, women, and children dying of malaria or starvation or sepsis matter. And what Kagan says about chickens can also apply to human beings.

A steak dinner or an afternoon in a fast car will probably do nothing to worsen climate change, but it *could* trigger a typhoon. If I eat meat without checking very carefully whether the farmer feeds antibiotics to healthy animals, I could be the extra customer who prompts that farmer to stock more animals, give them more drugs, and in doing so, enables a resistant bug to develop. "We shouldn't be thinking, 'I'm fine, I've had the [COVID-19] vaccine,'" says Sophie Harman, global health politics professor. "Before travelling, we should be asking, 'Why are you travelling? Who in those countries has had the vaccine?' Because this isn't just about the risk that others bring *to us*. We should be thinking, "Hang on, *I'm* the risk." And acting accordingly.[16]

Of course, almost everything we do *could* cause harm. Sometimes it does so much good, or is so necessary, and the risk is so small, that we still do it—and often that's OK. But at the very least, if there's a chance that what I do will kill someone, I have a moral reason *not* to do it.

Or think about this way: being part of harm can be morally problematic, even if you don't make a difference.

Consider this. It's 2021 and a healthy young man is deciding whether to get the COVID-19 vaccine. He knows it's still indeterminate just how bad the pandemic will be, how many vulnerable people will be exposed to COVID, and whether hospital beds will run out. He knows this depends on what the many people in his position will do. In other words, he knows he can *help* save vulnerable lives or *help* destroy them. He knows, too, that many people *are* getting the vaccine, wearing masks, avoiding social occasions, as part of this collectively crucial effort. If he doesn't do the same, he's free riding. Is this so clear-cut an outrage that he should *never* do it, whatever the cost to him? Not necessarily, but it gives him a moral reason to get the vaccine.[17]

There's another reason, too, instrumental but important. Sometimes, activism and communication can conflict with changing your own life. Both can use up time and money; it might be most effective for some people to fly around the world getting governments to change their tune on climate action. But, often, lifestyle change is part of promoting wider change. Rightly or wrongly, people are taken more seriously as agents of change if they change themselves (or decried as hypocrites if they don't).[18]

I talk to Brian Kateman, founder of the Reducetarian Foundation, which encourages people to cut down on meat and dairy. He tells me, smiling, about his vegan wedding. His mum was terrified. A couple of her friends were coming. What would they eat? In fact, they loved the food. One study found that New Zealand high school kids were more likely to stop littering if their friends did. Another study in the United States concluded that someone is 36 percent less likely to be a smoker if their friend has quit smoking, 34 percent less likely if a coworker in a small firm has done so, and 25 percent less likely if a sibling has quit.[19]

When I see other parents carting their kids around on cargo bikes, we exchange smiles. Often, we chat. In this world where corporations transcend borders and influence governments, we're all *citizen-consumers* now. We know that, and recognize each other as part of a movement, even if we don't think in precisely those terms.[20]

If even half or a quarter of the families on Mumsnet or Netmums went mostly veggie, or bought only ethical or secondhand clothes, that's not only a substantial drop in itself, it's also a serious message to corporations to think differently about what they do. By boycotting coal energy or turning down unnecessary antibiotics, we send a "signal" that we would accept a policy to require these changes. Even not having children can be (and often is) a political statement, albeit a complicated and potentially problematic one. Birthstrikers make a sacrifice more eloquent than words: a plea to decision-makers to stop and listen, because this is how bad things have got.[21]

OUR KIDS' MORAL INTEGRITY

One more thought. This is about the children themselves, and the global and ecological citizens they will grow up to be.

Daniel Butt, a philosopher at the University of Oxford, thinks even meat-eating parents should bring their kids up vegetarian. If he's right, then it seems obvious that parents should also bring children up so they aren't contributing to global emergencies. At least (and it's a big *at least*!) as far as possible. So is he right?[22]

The idea isn't that anyone could *blame* a child, or that they should blame themselves, for what they did when they were too young to be morally responsible. But it can still impact them, perhaps deeply, as they grow up into mature moral agents. Butt thinks that harming animals is (or could turn out to be) wrong in itself. I agree but, as with Kagan's chicken analogy, the argument applies even if it's not wrong. So long as they enjoy meat-laden diets or high-flying, materialistic lifestyles, our kids benefit from the suffering of people as well as animals. They free ride on the efforts of others to prevent it. They *help* do harm; they might just trigger serious catastrophe. They can later regret this, even if they had no choice.

We can feel what the philosopher Bernard Williams calls *agent-regret*, even for involuntary actions. The lorry driver who accidentally runs someone over regrets this in a way that separates him from any spectator, even a passenger sitting next to him in the cab. What's more, Williams says, this regret is not only understandable, it's also appropriate, even if undeserved. What might an adult feel, twenty years hence, whose childhood was spent on long-haul flights or whose education was paid for by conflict diamonds, even if none of this was his choice?[23]

To complicate things further, our children incur an additional, perhaps intolerable burden. They don't simply benefit from these harms: they become who they are because of others' suffering. This is physically so, Butt says, with eating animals. Their flesh becomes part of our children's bodies; they carry that legacy of harm with them, for the rest of their lives. But the same can apply to mental growth: If psychologist Cathrine Jansson-Boyd is right, teenagers plied with consumer goods can find their whole identity formed around materialism, before they are adults.

Perhaps you find these claims hard to accept. If so, there's still a more practical point to consider. If our children are used to flying regularly to the Maldives, eating beef burgers, or taking antibiotics at the slightest hint of illness, they may find it harder later to cut down on meat, stay local, or campaign for agricultural reform. If I'm right, we should raise kids who will work for collective change; we should also change our society, so they can tread more lightly on the earth, and on each other, than we have done. There is something very odd about doing this while fostering the very tastes that make things worse.

I'm vegan, but my daughters aren't. They aren't even vegetarian, though the older one has just turned pescatarian. Their diet has always been mostly veggie, and often vegan. But I haven't stopped them from eating meat at family and friends' houses. If I'd read Butt's paper when they were still babies, I probably would have done. And, as my little one has developed disturbingly carnivorous tastes, I wish I had. "I only want a meat one," she will say, suspiciously, when offered a hot dog. If she hadn't been given meat, she would never miss it.

I can't tell you exactly what to change in your family life. I can only help make moral sense of what you can do, in the bigger picture. But one thing stands out starkly clear, for both of us. If we care about our own children, or about other people, we'd better not be contributing to all these harms, *as well as* failing to change the current disastrous system. Being frequent fliers, drivers, meat-eaters, and not activists? That's the worst of both worlds.

Chapter 7 in Essence

Showing respect is crucial in day-to-day interactions—and it may be more complicated than you realize. Beyond that, family lifestyle changes matter, but they're not the priority. They're not the priority because this is a collective crisis, and these changes can be harder, inefficient, and sometimes ineffective, if the system is against us. The priority is changing that system.

Lifestyle changes matter because, in living as we do, we are contributing to harm. This applies especially to the way we travel (cars, planes), what we eat (meat, dairy), and how much stuff we buy. These changes matter because altering how we live is one way to get corporations, politicians, and those around us to pay attention.

Lifestyle changes also matter for our kids' moral integrity. They may regret being part of a way of life that destroyed their own future, and others', even though they had no say in the matter. More pragmatically, it will be harder for them to live differently, in a better future, if they have learned to depend on beef, planes, and consumerism for its own sake.

III HARD BUT NOT IMPOSSIBLE

8 MAKING DIFFICULT CHOICES

January 2021. It's 6:30 a.m., days into a new lockdown. I'm working on this book, bolted into the room my husband and I are taking turns to use as an office, while my five-year-old hammers at the door.

Schools and nurseries are shut. The house is carpeted with the remains of yesterday's craft, I already hate home schooling with a passion, I feel like I didn't sleep last night, and I've got classes to prepare for my university students. As a working mother, I've long felt like an imposter in every aspect of my life. Add a pandemic to the mix, and it's not so much that I'm trying to keep a dozen different balls in the air, as that I'm doing it on the edge of a cliff.

Pre-pandemic, 7–8 percent of parents in the United States qualified as suffering from burnout. During it, things only got worse. That's not just my own impression, borne of my experiences and those of my exhausted friends and colleagues. One study said parental stress in the United States "increased substantially." Another showed that UK parents' rates of stress and depression went up during the first lockdown in 2020, settled over the summer, then went back up again (with increased anxiety, too) during the second set of national restrictions. Things were particularly hard for those parents with kids under age eleven or neurodiverse kids; they were especially bad for mothers.[1]

If you feel like throwing this book out the window with frustration by now, I understand. If your kids are young, you've maybe been reading it between bouts of breastfeeding or catching up on work emails while lying on the floor in a darkened (but not too dark!) toddler's room. If they're older, you're most likely worrying about them: their grades, their friendships or relationships, their emotional health. Now here I am, piling new demands on your shoulders.

Parent activism "definitely takes huge amounts of time and energy," says researcher Lisa Howard. For those who are most committed, it involves planning meetings, online lobbying, and petitions. It means hours of organization, often at night when the kids are in bed. Howard describes a UK doctor she interviewed who was trying singlehandedly to "green" the NHS. This takes its toll.

So let's stop a minute, take a breath, and revisit the topic I discussed in chapter 2. Just how much is expected of me, or of you, in fulfilling our moral duties—including those to our own kids? Let's acknowledge the dilemmas we still face and see if the philosophical tools I've laid out can help find a way through. Then we'll talk about why it's *still* hard, to follow that path.

CONFLICTING CLAIMS

Time to check my privilege (again). I'm permanently tired, even out of lockdown. I'm emotionally drained. But I'm also one of the lucky ones.

My little girl, who yelled "Mummy" through a locked door in lockdown, has another parent. She had him then, too, waiting to make her porridge and cajole her into her clothes. We got through the pandemic by working in shifts. Not everyone had that option. The UK study I cited already confirmed what was, again, anecdotally obvious: things were much worse for single parents or low-income families. In South Africa, Xoli Fuyani works with families in settlements with no formal sanitation, high unemployment, shared taps, and communal toilets. "How are you supposed to parent," she says, "when you don't even have money to buy bread?"

Once again, it's worth spelling out what I'm *not* saying: that a mother struggling to feed her family, or in a women's shelter escaping a violent partner, should be donating to Oxfam or Black Lives Matter. That would be tone deaf to the point of callousness. Or that a parent caring round the clock for a child with severe disabilities, or a person in the worst of the "squeezed middle," juggling work, young children, and caring for their own bedridden mother or father, must create extra hours in the day to campaign for climate justice. Such parents may (or may not) have money to give, but time is a near-impossible ask.

Instead, I am reflecting on this for myself: how much I can and *should* do. In the process, I hope I can help other parents, in broadly similar situations, to figure it out for themselves. Because when it comes to us, there's a straightforward (if harsh) response to the claim: "But I've already got enough to do!" Being activists and raising them, or changing our lives, are not optional extras. They are not what philosophers call "supererogatory." They are essential to being good parents as well as being good global citizens, good ancestors as well as good intergenerational or ecological citizens. Plausibly, we owe our own kids more than we owe everyone else, but that's no moral "get out of jail free" card: not when all the responsibilities pull in the same direction.

What's more, we're not in normal circumstances. This is a crisis. "All of us who are in this are working flat out," says Maya Mailer, parent climate activist. "Other opportunities come up and I think: 'How could I do anything else?' because nothing else matters. This is my children. This is your children. It's all children. This is our planet."

Think collectively. Think of the sacrifices made in a war, or the less extreme but still dramatic changes we made in the pandemic. Not leaving our houses, except to exercise. Children not seeing grandparents or friends. Perhaps the significant costs I talked about in chapter 2 are too low a threshold for this kind of global catastrophe: perhaps we should do much, much more. Or think individually. Children are and should be their parents' priority. In emergencies, we *do* drop everything for them.

Kristina Stratton, Kharisma James, Marlene Cushni, Kelsi Wood, Jonathan Stevens. You probably haven't heard their names, but they all died in the last five years. Kristina Stratton and Marlene Cushni ran into burning houses. Kelsi Wood and Jonathan Stevens leapt into dangerous seas. Kharisma James jumped in front of a moving car. They all sacrificed themselves to save their children.[2]

Earlier, I asked you to imagine that your child was destined to contract a life-limiting disease at age fifty. Now imagine you can only save her by taking her halfway round the world when she's a baby, for an operation you must sell your home or get into debt to pay for. Wouldn't you do it? And wouldn't you *still* do a lot, even if there was only a significant risk that your child would get that disease, and your action only *might* stop it? Climate change is that disease. The next incarnation of COVID-19 is that disease. Pollution, institutionalized racism, violence against women, continued poverty, and antibiotic resistance are that disease.

Remember this, too: our children, for whom the future looms so desperately insecure, *didn't ask to be brought into the world*. Many of us made that decision for them, knowing what that world was like.

I find this reasoning compelling. But it's still not straight-forward. The overlapping demands of being a good global cit-izen and a good parent can clash with other things we *must* do, not only for the others in our lives, not only to pursue our own most central projects, but also for our own kids. What happens then?

"I try to set my own boundaries," says Maya Mailer, speak-ing from her London kitchen to my bedroom-and-lockdown-study, "but I've suffered from burnout. We all have." And this burnout can bad, not only for us, but also for our kids. When Robert Epstein and Shannon Fox examined survey data to find predictors for happy, healthy, successful kids, one of the top three was parents' ability to manage their own stress.[3]

Or meet Mr and Mrs Jellyby: modern-day versions of a minor character in *Bleak House*. They spend every hour of the day and almost every hour of the night marching, organiz-ing petitions and boycotts, or writing to politicians. They give all their income to charities promoting climate justice, social justice, or ending deforestation. To do all this, they ignore their own underfed, perpetually anxious, and uneducated kids. Parents of the year? Hardly, even if they are doing all this because they care about their children's future.[4]

In chapter 2, we asked what our children would prefer us to do, if they could reflect on this in terms of their lives as a whole. Let's ask this again, with the precarious state of the world in mind. They wouldn't want us to be the Jellybys, nor to be them themselves; but they might well want us to be *more* like that than we are now. In the face of global emergencies, I think my girls would rather I spend less time on them now (and certainly less money), so I can work with others to protect their future, and that they do the same, if necessary, for any children they have—especially if the alternative is leaving them unable to protect *their* children, at all. It's a gamble. I know that. These efforts others and I are making may *not* work. Still, I think my girls would want me to take that chance. I owe it to them to take it.[5]

But the devil is in the details. How *much* less time? How much less money? That depends, obviously, on how much our children need the things we would give up to change our lives, or become activists; on whether they are indispensable to a "decent life"; on how significant they are to the pursuits we value as families (and could sustainably *continue* to value as families).

In some ways, this is morally simple. Food, education? Essential. Innumerable overseas holidays? Not only inessential, but also impossible, in any kind of just and sustainable future.

In other ways, it's more complicated.

In drawing this line between parenting and parent *activism*, it helps to remember that both, if we are lucky, can be team games.

"I have a very supportive partner," says Maya Mailer, "and that allows me to do a lot of this." The same goes, she says, for many parent activists. It doesn't tend to be *both* parents, throwing themselves heart and soul into the movement at the same time. While I was writing this book, my daughters had a stressed, distracted, and often absent mother. That weighs heavily on me. But I couldn't have written it at all, not while holding a full-time job and living through a pandemic, if I hadn't had the support of my husband, friends, parents, and in-laws. In this, too, I recognize my privilege. For single parents or those without family or affordable help, the choices are harder still.

Parenting and activism are also long games. In making sustainable lifestyle choices, with an eye on their future, I must often put my children at odds with what their peer group does, and I must do this at an age where fitting in seems all important. I get enough complaints already for sending them to school with snacks in reusable boxes, when (they claim) their friends have plastic-wrapped treats. I ignore these. But what do I do if my daughter, as a teenager, is invited to join friends on a coveted overseas holiday, complete with flight, or gets the opportunity for once-in-a-lifetime work experience—in Australia?

If we focus on changing the baseline—on altering how we live day to day and year to year, equipping our children for a just and sustainable future—then exceptions can be more easily weighed as they arise, and not necessarily agonized over. We gave up our car, a few years back. With the cargo bike, we've barely missed it. But we borrow or hire a car (preferably electric) from time to time. I haven't flown for years, but that doesn't mean I'd never make an exception, or insist that my children never did. So long as the reason for flying was important enough, and it didn't become the "thin end of the wedge," opening the door to more and higher carbon choices.

However, some of these quandaries *are* unavoidably long-term. Take the choice of whether to use private schooling: a tough one for those who are both ethically minded and privileged enough to have the option. It's an unnecessary luxury, on the face of it: a huge investment in fees that could, instead, do considerable good in the world. But what if the local schools can't provide an adequate education, either because they are poorly resourced or managed or because they can't support your child's special educational needs? And what if your child is being horribly bullied where they are, and finding an alternative is critical?

The philosopher Adam Swift points to the difference between going private under those circumstances and sending your kids to a vastly expensive school when other good-enough options are available. It's the difference between removing them from a situation that would harm them and giving them a head start over everyone else's children; between protecting them by temporarily bypassing a messed-up system that you want to change, and actively seeking to perpetuate that system.[6]

Of course, private schools can give our children things worth having in themselves: extra chances to cultivate an informed love of music, sport, or literature. Maybe even the kind of engagement with nature, by planting trees or playing in green spaces, that I lauded in chapter 6. Private schools, after all, can *afford* just such green spaces. These experiences are deeply valuable: passports, perhaps, to deep-rooted human joy. But in making their decision, those parents with the privilege to do so must ask what exactly they would be paying for. Is there no cheaper way to introduce children to cultural and natural delights? Would state-funded schools really give them none of these things, or only less of them (and, perhaps, a whole other set of valuable insights and experiences, instead)? Is this about what your child *needs*? Or is it, again, about relative advantage?

Relative advantage can matter. We saw that in chapter 2. But not to an unlimited extent. The norms of a society can, in part, dictate what is needed for our kids to flourish. Some philosophers allow that to pull against the needs of the vulnerable. But with so much at stake—for the global poor, for future generations, and *for your own child's future*—that pull looks morally weaker. In raising my girls to be advocates for global change, I may not be preparing them to succeed optimally as our society currently recognizes success, but that society needs to change, for all our children's sake.

Or put it like this. If you are paying tens of thousands a year simply to jump your child up the social or economic queue, are you really investing in the future you want them to have? Or are you just buying them a more comfortable deckchair on a sinking ship, which might, instead, be made seaworthy?

IT'S NOT ALL SACRIFICE

I've highlighted the hard choices, here, because there is no editing them away. But the choices I've talked about aren't *all* sacrifices. Far from it, even if some of them feel that way at the time.

Take the lifestyle changes that help protect our children's futures. Brian Kateman, who runs the Reducetarian Foundation, is both refreshingly honest, and encouraging: "Cutting back on animal products can be difficult. There can be this idea in advocacy it's easy, and it can be easy, but it can be hard. But . . . it's wonderful when you find yourself enjoying all these types of food, when you start to feel lighter and more energetic, like you are doing something good for yourself and the world. It's quite an addictive transition."

Listening to him, I reflect on my own journey to veganism. I was daunted by the practicalities. How would I cope without cheese? Would I ever be able to eat at friends' houses again?

In fact, it took a bit of adaptation, some new basics in the cupboard, a few awkward conversations, and some minor grumbling from my husband before he threw himself into the new culinary challenge (and found it, he tells me, both easier and tastier than he had expected). Then it was fine. Better than fine. Friends and family were accommodating—and plenty of them have cut down on meat and dairy too. I *do* miss cheese, but nowhere near as much as I feared. And, like Kateman, I feel physically and mentally better. The science backs us up. Eating more plants and less red meat means less risk of cancer and heart disease.[7]

Then there's travel. I flew plenty, before I thought deeply about these things. I loved to see the world. And, yes, it makes me sad that my kids can't take that privilege for granted. But for many years, our holidays have been almost exclusively on this cluster of islands we were born on. Exploring Cambridge, York, Winchester. Climbing sea cliffs or lying on my back in the Cornish sun. Hanging onto my squirming daughters as they learn to bodyboard in the Irish sea. Those trips, too, have been wonderful and relaxing, illuminating and exhilarating.

Lockdown meant many privations but, for me, even that was a valuable lesson in the importance of homely pleasures:

local walks I never knew existed, burgeoning wildlife, snuggling up with my daughters, reading stories of adventure.

Active travel—walking and cycling, rather than driving—is good for physical *and* mental health. The polluted school run isn't always a classic "tragedy of the commons" (where it's individually rational to drive but collectively rational not to). It's crucial to campaign for low traffic zones or bike buses, but even while we're waiting for things to improve, the individual health benefits of walking and cycling can be greater than the additional risks from air pollution. At least in all but the most polluted areas.[8]

Activism, too, can involve adjusting the things we do together as a family, not making the experiences worse. It can even make them better.

Children need quality time with their parents. That could be time on a march for social or global justice, time planting trees together, or time in a theme park. Our lives are no less rewarding for making it more of the first two, and less of the latter.

Just as being brought up too materialistic can warp our children's potential, so, on the flipside, can doing more for others *benefit* their mental health. Psychologists in the United States found not only that happiness and "ecologically responsible behavior" go together, but also that the two are linked by "intrinsic" rather than "extrinsic" personal values: growth, relationships, and community involvement rather than popularity, image, or financial success. A 2020 Mexican study showed that children are more altruistic, more frugal, fairer, *and happier* if they are more connected with and concerned about the environment.[9]

Burnout helps no one. But for parents, as for children, some activism is good for mental health. "People are anxious because they don't feel things are being done to address the issue," Susan Clayton, psychology professor and fellow of the American Psychological Association, reminds me. Ugandan activist Herbert Murungi tells me: "I cope with the anxiety by associating with other people who value my work and reciprocate. I'm happy that the next generation will know I stood up for a cause bigger than the material world."

"I'm living my purpose," says Xoli Fuyani, who wears an incredible number of hats, "and what I do is me. When I take groups of kids hiking I don't see that as a chore, it's a time for me to connect with them, with nature, with being out. Approaching it from that perspective, what I do doesn't feel like a lot."

THE PSYCHOLOGICAL BARRIERS

So far, (I hope) so morally helpful. But in practice, even when our choices are philosophically clear cut, we don't act as though they are.

According to Yale scientists, only 33 percent of US adults are "Alarmed" about climate change, believe it's happening, we're causing it and it's very bad, and support strong policies. Nine percent of adults are "Dismissive." They actively deny that climate change is happening, is dangerous, or is caused by humans. Often, they think it's an elaborate hoax. Among the rest, 10 percent don't think it's a major risk, 5 percent know little about it, 17 percent haven't made up their minds, and 25 percent think it's a serious threat, but comparatively low priority. However, even those who *are* alarmed don't always do anything about it. To quote a Londoner interviewed by psychotherapist Robert Tollemache: "I'm not living the life, I'm not walking the talk." Even as acquaintances died and doctor screamed warnings on social media, some people chose to deny the reality of the pandemic.[10]

So let's delve deeper.

Time matters, when it comes to motivation.

"There must come a point at which generations in the future become increasingly abstract to you, and you don't . . . care about them in the same way." So said another of the Londoners interviewed by Tollemache, who was trying to find out why they behaved in environmentally unfriendly ways. "Although there are large individual and cultural differences in the ability to delay gratification," says Norwegian psychologist and politician Per Espen Stoknes, "people in modern societies still overwhelmingly weigh present outcomes as much more important than distant ones."[11]

Geography *seems* to matter: most of us would do a lot to save our neighbor's child from going blind, if we could, but we throw away flyers asking us to donate $20 to save the eyes of a child on the other side of the world. But when a philosopher and a psychologist put this to the test, using carefully designed scenarios, it turned out not to be distance that made the difference, but other factors that often go along with it: how directly the information gets to the person concerned, shared group membership, and efficaciousness (for example, whether there is another potential helper closer to the person in need).[12]

Our own embedded attitudes and tastes also matter.

I love chocolate. I *really* love it. It would be the last culinary luxury I would give up. I can want (and demand) a world where chocolate comes with a cleaner moral slate: where it is produced without cutting down rainforests or forcing farmers into destitution. That's easy enough for me, since I can afford chocolate that doesn't do those things. But campaign for a world where much less chocolate is eaten, even if that would mean giving up my own favorite luxury? That's much harder. So I mostly try not to think about it.

I'm not alone. We want our values, the attitudes that determine how we act, and the facts we accept as true to fit nicely together. When they don't, we must alter one or the other. Otherwise, we're in a state known by psychologists as "cognitive dissonance." And, unfortunately, it's not always our behavior that changes.[13]

Smoking causes cancer. But rather than quit when presented with this unpleasant piece of information, smokers tend to do one of four things. They adjust their view of reality ("I don't smoke all that much, not compared with the man next door"). They decide they can "cancel it out" by doing something else that's healthy, like exercise. They underplay the worry, questioning the scientific evidence or appealing to anecdotal examples. Or they deny the science outright, usually with some sinister explanation of manipulation by scientists. Climate or COVID-19 denialists do the same.

This is about individual psychology, but it's also more insidious than that. It's about how society pushes us to ignore global emergencies. "Bob," who cares only about himself, his own children, and their material well-being, is in many ways the natural product of societies like ours.

If denialists want "evidence" to support them, they need only go online; in this so-called post-truth world they won't need to look far. Vested interests (a terrifying, institutionalized denial movement) fall over themselves to provide it. But there are also more subtle factors at play. There are no political denialists when it comes to antibiotic resistance, says Otto Cars. They know it's a problem. But still collective action doesn't happen. It doesn't even make it onto most people's "worry list."[14]

This is about who we are or are socialized into believing ourselves to be. In the early 2000s, sociologist Kari Norgaard studied a Norwegian village highly vulnerable to climate change. The community knew the facts. The ice was melting beneath their feet. The economic impact, too, was real. But they carried on as though nothing was happening. They did it because the facts were incompatible with their culturally created image of reality, their individual identities, and their sense of empowerment.[15]

We have form, on this. Clive Hamilton, professor of public ethics, compares climate denial or apathy to resistance to Einstein's Theory of Relativity, which threatened existing theoretical commitments, or to conservative papers accusing Winston Churchill of "alarmism" when he warned in 1934 that Hitler was a threat to world peace. Older, conservative white men are significantly more likely than other Americans to deny climate change. Their cultural identity is most bound up in the status quo.[16]

This is about how we relate to nonhumans. In the Global North, we see consumption itself as a marker of success; we're constantly afraid of losing our competitive edge. In the process, as we saw in chapter 3, we've come to treat the rest of the world as a commodity, not something we are part of and depend on. As psychoanalyst Sally Weintrobe puts it, we used to love nature, but capitalist culture erodes that. "Over centuries," says

Nadine Andrews, chair of the Scottish Climate Psychology Alliance, "we've come to this very toxic set of beliefs about our separateness and superiority, which has led to this situation of denying that there are physical limits. We think those limits don't apply to us because we are superior and ingenious."[17]

Humans *are* ingenious, she says, but not ingenious enough to get away with this collective denial of reality. "Pushing wild animals out of their habitats and into closer contact with other species and humans makes pandemics more likely. COVID is feedback. As others have also said, it is telling us that our relationship with the natural world is out of whack. We need to do something about that relationship. Instead, we shoot the messenger. We treat the symptoms."

The short version? Denial of our dependence on the Earth is a "modern neurosis." And it comes at a price.[18]

This is also about the people around us. In 2008, a mid-priced hotel in the American Southwest randomly assigned different messages to 190 rooms, to encourage guests to reuse towels. One message appealed directly to the environmental benefits. The other said that most other guests participated in the scheme. The second sign was significantly more effective.[19]

As Stoknes puts it, "We are, at our core, imitators." Not only do we try to make facts fit our attitudes, we also are predisposed to have the same attitudes as our friends. This may be a matter of grabbing at a moral "get out clause": the idea that this is someone else's job too, and they aren't doing it. Or it may be about embarrassment. The Londoners interviewed by Tollemache felt pressure to conform. Some were afraid of being seen as "preaching" to friends and family. Some ridiculed or disparaged activists themselves.[20]

I'm forty-five now, and less bothered about coming across as either curmudgeonly or moralistic. Even so, I don't always speak up. It's awkward to break up the flow of assumptions, the excited small talk, with a dose of reality. The guest who asks, "Where did that meat come from?" or brings up COVID statistics to a happy group of unmasked friends might hope to make a difference; she'll also worry about being kicked off the guest list next time. It's also disheartening, feeling like no one else cares.

But here's the ultimate bleak irony: we might *all* be scared as hell for our children, and too embarrassed to speak out. According to Lichtenberg, we often interpret situations in line with others' interpretations, or what we think of those as being. "We think 'Oh that sounds pretty bad,'" Susan Clayton tells me, "but when we look around, we think that no one else seems that worried and so we don't express alarm." We don't realize, she says, "that other people are taking their cues from us as well and interpreting our lack of expressed alarm as confirmation that they don't need to worry." The problem becomes mutually reinforcing: a collective failure to do what we all, perhaps, would prefer.

PSYCHOLOGY IS NOT MORALITY

These psychological barriers help explain our apathy in the face of intergenerational disaster. But (and let's be very clear on this!) they don't make it OK.

As we move away from paradigm cases of individual responsibility ("Jack stole Jill's bike") toward situations like climate change, something changes in the collective mindset. "The view that morality is involved is weaker . . . ," writes the philosopher

Dale Jamieson, "perhaps disappearing altogether for some people." But morality doesn't really disappear, just because it suits us to think it does. (Suppose Jack took Jill's bike because he felt pressure to fit in with cycling friends, and persuaded himself it wasn't that bad. Motivation? Yes. Justification? No.) We share responsibility when our actions combine predictably to cause suffering, or we are part of harmful ways of life. We *should* organize to prevent suffering, even if we didn't cause it. We have duties to our own children that can only be fulfilled by changing the way we live together. We've seen all that already.[21]

Some psychological obstacles amount to blocks. Leaving the house may be impossible for an agoraphobe, so they can't be expected to do it. As philosophers put it, "ought implies can." But for most of us, most of the time, that's not the case. It's not easy to overcome these psychological hurdles, but it's not impossible. They are difficulties to be chalked up, along with the other costs of time and money and effort, in determining just how much can be expected of us. They're not the end of the discussion.

Suppose I don't sign up to help at my kids' school fair. I'm not too busy, but I really hate making small talk all day. In this case, the stakes are relatively low, and crowds really do make me miserable, so maybe I'm off the hook—especially if I buy lots of raffle tickets instead. But sometimes the stakes are higher. Perhaps we need thousands of signatures for a petition to stop the school being closed altogether and I'm asked to help collect them. My excuse—"But I don't like talking to people!"—doesn't cut so much ice then. Not with my daughters' education at stake.

Now, it's our kids' whole future at stake. I can't absolve myself from responsibility, on the grounds that I'm trapped by socially imposed psychology. It's easy just to say that we are short-term thinkers, but perhaps we *shouldn't* be: not when future generations are humans too, and there's a huge moral difference between economic "discounting" and accepting policies that would destroy our great-grandchildren's lives. Instead, we must figure out how to get ourselves un-trapped and move forward.[22]

Chapter 8 in Essence

If I'm right, we owe it to our own kids (and everyone else) to be activists and raise them, and to live differently. Being this kind of good parent might mean not doing other valuable things for our own children, ourselves, or other people we love. It's not all sacrifice—quite the reverse—but there are some tough choices. You'll have to make them for yourself, but I can offer some moral tools, to help.

One is this: Think long-term, ring-fencing what matters rather than obsessing over each individual choice. Another is this: To find that balance, ask yourself what your child really needs and what you need as a family. Are you cultivating tastes that have no place in a sustainable future? Are you giving your kids a fair chance of a decent life, or buying them a competitive advantage on a sinking ship? Ask what your children would rather you focused on if they could see this in terms of their lives as a whole.

Then there are the psychological obstacles. We're socially wired for short-term, consumer-driven thinking. We're predisposed to deny inconvenient facts rather than change our attitudes or behavior. But these hurdles are explanations, not moral justifications. We must turn to psychology itself, for help in overcoming them.

9 A PSYCHOLOGIST'S TOOLKIT

The Russian assault on Ukraine continues as I write this. Atrocity after atrocity is revealed, day on day. India and Pakistan have been hit by unprecedented heatwaves, their citizens struggling with temperatures literally too hot for human bodies to survive. And the headlines flashing up on my phone? Kim Kardashian's gratitude to her ex-husband for retrieving a sex tape from another ex. A couple of months ago, I watched the film *Don't Look Up!* I laughed, even while I wanted to cry. But reality is worse—and less funny—than fiction.[1]

I see ads everywhere: images of the "good life" with material luxury front and center. My children plead for more toys when they have plenty already. I swerve to avoid the charity collector on the street. We call them "chuggers" here, short for "charity muggers," our shame, no doubt, fueling resentment. My government talks about "net zero" but without any real plan of how to get there. I read the new IPCC report, and I think how can this be real, when all around me are carrying on as usual?

I need to change how I think—or change it enough to act differently. But, of course, it's not that simple. These traits are deep-rooted, and the shift we need to make isn't like gearing up for some one-off act: a single sacrifice that will save the world, like the father flying his plane into the alien spaceship at the end of *Independence Day*. I tell myself, firmly, that if psychology is priming me to do what's morally wrong, then so much the worse for psychology. I just have to fight it. But, unsurprisingly, the shackles remain.[2]

"[They] believe they ought to do what they do," says philosopher Judith Lichtenberg, of people who dedicate themselves to the environment or empowering the global poor, "but they also want to, because their acts affirm the kinds of people they are and want to be and the kind of world they want to exist." Finding consistent motivation is about identity and self-image. It's about finding fulfillment and motivation away from short-term, narrowly defined self-interest and consumerism.[3]

Over the past year or so I have read as much as I can, from psychology and psychotherapy to philosophy, anthropology, geology, and literature. I've interviewed activists and climate and eco-psychologists. From this incredible and sometimes bewildering mass of information and advice, I've pulled out the truths and strategies that I have found most helpful, in finding a way forward.

FACING EMOTIONS

Worry. Fear. Grief. Anger at politicians and corporate leaders. Frustration. Shame. Resentment that I must give so much of my time and mental energy to fight something so completely avoidable, and that if I don't, my daughters' generation will bear even more of the brunt.

That's what I feel, now, when I open my eyes to reality.

Climate scientists, too, are afraid. I read their handwritten letters on the website Is This How You Feel? and I collect adjectives. They feel concerned, scared, anxious, apprehensive, worried, and fearful. They feel sad, depressed, discouraged, distressed, upset, dismayed, and despairing. They feel ashamed, guilty, outraged, exasperated, disappointed, angry, disgusted, annoyed, frustrated, and infuriated. They feel helpless, powerless, overwhelmed, apathetic, tired, perplexed, astonished, bewildered, bemused, confused, and dumbfounded. They feel "like nobody's listening."[4]

This fear, anger, and palpable sense of loss is not only, or not primarily, for themselves. It is for the vanishing systems, the abused world, its abused people, and their own children and grandchildren.

These feelings are hard. It's no wonder we don't want to open this particular Pandora's Box. But what's the alternative, for us or for our kids? Do we engage in flat-out denial, living in a fantasy world while undermining our children's future? Or do we endure the constant mental aerobatics of suppressed rage, suppressed grief, and spiraling anxiety?

"I put up a wall," says Charles, a retired medical researcher interviewed by Tollemache, explaining his ability to live in the way he does, despite the reality of environmental harm. In *Great Expectations*, the hero, Pip, is set up in life by a mysterious benefactor. Years later, this turns out to be the violent convict Magwitch. Everything Pip believed about himself seems destroyed. Charles is all of us. I think we know that already. But according to psychotherapist Rosemary Randall, Pip is all of us, too. "In the eighteenth century," she says, "only the minority tasted slavery each time they drank a cup of sugar-sweetened chocolate. Little has changed today. People rarely make these connections, the relationships are not apparent to them, and their reaction to being shown them is often similar to that of Pip when confronted with Magwitch . . . shock and angry disbelief."[5]

It's an uncomfortable analogy, but it makes psychological sense. In the face of unpleasant reality, there are three forms of denial. There's full-on denial. There's negation or "saying that something that is, is not": often a first response to shock, or painful news. And there's disavowal. This is longer lasting than negation and peculiarly insidious because it involves trying simultaneously to know something, and not know it. Call it "double-think." Call it "perverse thinking." Either way, we avoid reality, more and more systematically. We find ever-cleverer ways of tricking ourselves. As we do, the suppressed anxiety grows. Our response? More disavowal—and therefore more anxiety.[6]

If a difficult experience can't be contained, warns psychologist Paul Hoggett, we must get rid of it, or it'll stick in the system like undigestible food. Perhaps our internal conflict shows itself in an attitude of entitlement. Perhaps we project suppressed emotions by blaming immigrants or the global poor. Perhaps we act blindly, without thought, just to be doing *something*. Perhaps the anxiety becomes overwhelming: a kind of paralyzing despair, so we do nothing at all. Perhaps, if we're privileged (and, I'm guessing, don't have kids), we become fatalists, or what Mary Annaïse Heglar brilliantly calls "doomer dudes." We claim, almost as falsely and dangerously as denial-ists, that nothing can be done.[7]

Of the six groups of Americans identified by Yale scientists, only those "Alarmed" by climate change and those "Dismissive" of it, at the opposite ends of the scale, are even roughly consis-tent in what they think, feel, and do. That's what psychologist Per Espen Stoknes thinks. That leaves 58 percent of US adults doing constant mental gymnastics. And *that*'s just about cli-mate change.[8]

However difficult to live with, my emotions, and those of the scientists on Is This How You Feel?, are appropriate responses to a terrible situation. Climate change is terrifying. Pandemics are terrifying. I *should* be angry that young women can't walk home without keys between their knuckles and young black men are afraid of being shot by the police. On a video call, Edinburgh to Uppsala, I ask Otto Cars, infectious diseases expert, if we should be "very worried" about antibiotic resistance. "Yes," he says. "Definitely."

So our first psychological step, easier said than done, is to acknowledge difficult emotions, work with them, and work through them. "The work involves facing our self-idealizations, mourning our illusions and bearing difficult feelings," says psy-choanalyst Sally Weintrobe. For many of us, it means admit-ting shared responsibility or collusion, and learning to live with that. It takes introspection, it takes patience, and it takes courage.[9]

Stoknes describes his reaction to climate change as a "Great Grief," hitting him "like a fog." But in this grief, he says, there is a way forward. Through depression, we come to understand that we are not the masters of the universe. One chapter of his book is titled "Stand Up for Your Depression!" We come through despair, he says, to find less destructive ways of living: even a kind of hope.[10]

This is not the buck-passing version that Greta Thunberg has in mind when she tells older generations that she doesn't want their hope. It is no passive optimism, sitting back and assuming someone else will sort it out. It's certainly not the kind of optimism shown by some of Tollemache's Londoners, which amounted to feeling entitled to do whatever they wanted. We might call it "dark optimism," as Hoggett does. Or "active skepticism," like Stoknes: hope through determination, in a desperate situation.[11]

"Are you hopeful?" I ask South African parent activist, Xoli Fuyani, toward the end of our conversation. She tells me: "Not always. But if I look at the work we have done and what the young people are doing, I do feel hopeful." I find this glimmer of light, too, in the letters from climate scientists: a faith in human ingenuity, an unwillingness to give up, a resolution to help.

"Humans," says Anna Harper, research fellow at the UK's University of Exeter, "are remarkably resilient, creative, and intelligent. . . . Now is the time to apply all of our creativity and innovation to the big problem. . . . So I am *hopeful*, in the end." "Ultimately," writes professor and marine ecologist Greeta Pecl, "I do the research and communication work that I do as I just want to be able to look future generations, including my own children, in the eye. . . . Although it's tempting, I won't let myself lose hope because a future where we give up is just too horrible to contemplate. I won't do that to my kids. Or yours."[12]

Putting a moral slant on it, I would call this *earned* hope.

There is much we can each do, to get through this process. We can find solace and inspiration in art, film, music, and literature: in the turning of a lens, tragic, satirical, or tragicomic, on our collective wrong-headedness; in fictional blueprints for a better future, laid out for us by authors like Kim Stanley Robinson. We can seek professional help, psychological or psychoanalytic. But some psychologists think the shift we need is beyond the powers of adaptation of any single individual; so much of who you are, they say, depends on who you are with.[13]

There's a positive slant on this: the power of community.

"There was this sense of hope and solidarity," says Howard, of her interviews with parent activists. "That was hugely social, about seeing and connecting with other people and seeing others being influenced by them. A real sense of, 'Wow I have power in the face of these monolithic governments and COPs and big corporations!'" "It's awesome," says Shugarman, "that we can be part of creating a better future."

Mid-pandemic, exhausted from homeschooling and teaching online, the last thing I felt like was another video call, with parent climate activists. (Full disclosure: I wanted to wrap myself in a giant blanket, drink gin, and watch *Bridgerton*.) But as soon as I joined the calls, I was energized. I could call it a kind of hope that I find in engaging with these incredible, determined people and trying to play my own small part. I could call it solidarity, comradeship, or sheer shared dogged determination. It's probably a bit of all of them. Whatever it is, I know that I need this.

There's also a negative slant: an obvious, if disturbing, worry. It's disorienting and unmotivating to be around people who don't seem to care. We've seen that already. As Stoknes asks, "Do I have to change my friends?" If we must, then this psychological overhaul comes at a terrible cost.[14]

Hopefully, we don't. For one thing, the chances are that most of your friends, like most of mine, aren't in the 9 percent who are "Dismissive" about climate change or the equivalent when it comes to antibiotic resistance, pandemics, racism, sexism, global security, or global poverty. If psychology professor Susan Clayton is right, this is a self-perpetuating problem, where our mutual fear of speaking out keeps a whole group from acting. So there's an obvious answer: speak out and go on this journey together.

Furthermore, we can make new friends without ditching the old ones. Leanne Brummell tells me how she started her own action group. "I would sit on the street knitting and other local parents would say, 'What are you doing, Leanne?' and I'd say, 'I'm knitting,' and then I'd tell them about the gas and how that's coming closer." She grins. "They all think I'm pretty strange now." But, she says, she's encouraged and supported by the other parents in the Australian and global activist movement. She's even inspired by her 5:00 a.m. video calls.

It's not plain sailing, even when we have our "people." Collaborating brings its own emotional and mental challenges. We must find ways of building visions together, actively supporting each other, developing consensus when we won't agree on everything. But at least we'll be doing it together.[15]

CONNECTING ACROSS TIME AND SPACE . . .

While I'm writing this, visitors are experiencing an innovative exhibition in Wolverhampton, a UK Midlands town with a long industrial history. It's called *Dear Tomorrow* and it features a three-meter diameter globe, lit pale and glowing, a soundscape, and letters, videos, and photos from across the world. These are drawn from the website of the same name, the brainchild of parent-activist Jill Kubit and behavioral scientist Trisha Shrum. They make heartrending, but motivating, viewing. Parents write to their children or future grandchildren. Young people leave messages for future generations. Some speak to themselves in 2050, to their own possible children, or just to "the future." Visitors are encouraged to reflect, to write their own messages for a further exhibition later in the year.[16]

"When you're reading this," one California mother writes to her son, "things will probably be a lot scarier and more dire than they are, even now. . . . Almost every year that you've been alive you've seen California's fire season and watched the sun and sky get red and felt ash rain down from the sky, it was never like that before, even just a few years before you were born."

"Dear future family . . . ," writes an anonymous contributor, in December 2021, "I hope the time between this moment . . . and the moment you're reading this letter is a time you can look back on as an era of ecologic conservation and transformation rather than of tragedy. Is that too much to hope for? Here's what I am doing now to ensure those hopes for you."

These letter writers know what I know. They know that the victims of global crises include the children standing beside us, holding our hands to cross the road, as the cars get faster. They know this is a psychological kick in the pants like nothing else can be. "If you have children," says Harriet Shugarman, "how can you say their future is doomed?" Xoli Fuyani was an activist well before she was a mother, provoked by the inequalities and segregation around her. But, she says, "My awareness of the urgency of climate change changed, when I became a parent." Mine did too: it became more visceral.

Like Fuyani and Shugarman, I feel this instinctively. But Susan Clayton agrees, with years of psychological research at her back: "Getting parents to think about their children and their well-being and future does seem to be an effective way forward. I absolutely do think it's a powerful source of motivation."

Promising words. But if chapter 3 is right—if philosophers like Rupert Read are right—we need to be good ancestors, to be truly good parents. Casting our minds and hearts thirty years into the future is only a start. To avert environmental crises, dominant, consumer-driven societies could have learned from Indigenous communities, already attuned to long-term thinking. An Iroquois principle, known widely as "Seven Generations," requires communities to make decisions with wisdom gained from past experiences, and consider the fate of those still unborn.[17]

How do we get from here to there? How do we learn to motivate ourselves by what things will look like many generations hence, when it's a jump even to take seriously what our children and their contemporaries will face? We need, says Read, to cultivate "a *deep* care" for our children, extending across the future and the nonhuman world.[18]

We might start with deep time thinking.

In Dorset, where my dad grew up and my parents live now, I watch my father and husband boulder-hopping with my bigger girl; my smaller daughter peers into rockpools, holding my mum's hand. Thirty-odd miles from where we are, my granddad worked for decades as a vet, out at all hours, this rolling, half-tamed county becoming part of his blood. My little one, another animal lover, stops to pat every dog we see. She spots wildlife even before her birdwatcher dad, who inherited this love, in turn, from *his* grandfather.

I am connected to past and future here: two generations back, one forward. But I am situated in them more deeply, too, if I can only see it. Some six thousand years ago, a neolithic enclosure was built at Maiden Castle, another of our favorite picnic sites. In the early Iron Age, it became a hillfort; nearly two thousand years ago, a Roman temple complex stood there. My daughters and nephews roll laughing down the steep grass embankments behind which once lay battle-scarred bones, thousands of years old.[19]

Or deeper again. My uncle is a geologist; we are a family of climbers. I ought to be used to seeing a place through its rocks, and here they tell a dramatic story. Marine chalks deposited over millennia, on mudstones, sandstones, limestone, and clay. They are prone to upheaving themselves, these Jurassic legacies, overturning the landscape with startling immediacy, churning up fossilized scraps of the distant past. We have scrambled along the boulders to find dinosaur footprints: literal imprints of a history so deep in time that I could barely appreciate it, even as I tried to impress it on my daughters and nephews.

This is my attempt at deep time thinking, or what geologist Marcia Bjornerud calls "timefulness": stretching my awareness of past and future across geological timescales. She tells us that geology is a lens through which we can see the Earth's history, and our place in it. She reminds us that we are in the fourth atmosphere our planet has experienced, and that that there have been at least five mass extinctions and many smaller ones, even since the start of the Cambrian era, around 540 million years ago. She tells us that carbon dioxide helped kill off the dinosaurs. We would do well to remember that, as we live through a sixth mass extinction, amid increasing greenhouse gases and rising temperatures, in an age scientists call the Anthropocene.[20]

That's not easy, but there are strategies to help. Imagine our cities and landscapes, far into the past, advises anthropologist Vincent Ialenti. Then try to envisage them far in the future. Think how our day-to-day artefacts would look, displayed in a museum hundreds of years hence, or what the archaeologists and geologists of the future will uncover. I read another beautiful, disturbing book, by my colleague David Farrier, and picture layers of plastic, folding into the ocean floor; the billions of fossilized chicken bones; the sim cards and vending machines. I think of Edinburgh, the city my children were born in, built on dead volcanoes. I think of its edges, some of its most vibrant communities, disappearing under water. I think of Miami and New Orleans, Dhaka and Venice and Shanghai as Farrier describes them, settling further and further down, under the sea, becoming just another stripe in geological history.[21]

Does this help? It just might, because it reminds us that the physical world we take for granted, even the air we breathe, is not permanent. When change comes, Bjornerud warns, "It can be breathtakingly sudden. Tinkering with atmospheric chemistry is a dangerous business; ungovernable forces can come out of thin air." While I can't do much about fifty million years hence, I can—we can—about the next thousand.[22]

To be good parents and global citizens, we need to see not only across time, but also across continents: across people. I find myself wondering if the same tools could help. I tread with caution, here, knowing that some psychologists are wary of empathy, which is disturbingly inconsistent. I know parents and children coping with desperate circumstances don't need well-off white women making it "all about us," by public tears at photos like Amal Hussain's. They need the world to change.[23]

But to make it change, we need to change ourselves. An environmentally virtuous person, says the philosopher Dale Jamieson, "would appreciate the consequences of her actions that are remote in time and space. She would see herself as taking on the moral weight of production and disposal when she purchases an article of clothing." Suppose, when I looked at my beautiful city with its tidy streets and safe green play parks, I taught myself to see the other side of this picture: children sorting rubbish in vast, hazardous sites in Accra or Mumbai; miles on miles of plastic bottles, clothes, and broken electronics, lifted and turned in small hands. Might that help me to be "mindful," before buying another plastic toy? Might it help me to think carefully about the kind of corporations and politicians I want to support?[24]

I don't know. But perhaps it's worth a try.

. . . AND SPECIES

"What is this life," asks the poet William Henry Davis, "if, full of care / We have no time to stand and stare. . . . No time to see, in broad daylight / Streams full of stars, like skies at night."

What is this life—but also, what are *we*—so long as we are blind to the beauty, strength, and vulnerability all about us?[25]

Many global crises arise from collective disconnect with the nonhuman world; if we're to address them, we need *individual* reconnect. Dale Jamieson is no virtue ethicist, but he thinks the way to break the psychological duck, when it comes to climate action, is to cultivate environmental virtues, like respect for nature, or his version of mindfulness. For Xoli Fuyani, environmental educator, the bond with nature is "where it starts."

I turn to Nadine Andrews, psychosocial researcher and nature-based coach, for advice on rebuilding this bond. Among (many!) other things, she is a qualified mountain leader. But, she says, this isn't just about epic treks or iconic views at sunrise. They help, but to make this connection part of us and our children, rather than a one-off experience, we need to address what she calls the "culture of dysattention." The endless urge to switch tabs on a browser, check social media (yes, I've done it a dozen times, just in writing this section), or read headlines while you're on a walk with the kids.

We need to abandon our props, day by day. Andrews leads night walks near the full moon, so there's no need for torches. But there is wonder, too, she says in the plants and animals in your garden or the local park; in watching how they change, over the seasons, the weeks, even over the day. I watch my six-year-old crouching by a pond, fascinated by the wriggling tadpoles, the water snails clinging to tangled stems. I find myself reaching for my phone, to take a photo, share the moment with family. But my little girl has it right. *She's* in the moment. Children unlearn that skill, Andrews says. They need to keep it, and we need to relearn it. We need to start noticing.

When my husband can tear himself from *his* phone, every walk has a whole layer more of experience for him than for me, because he is a birdwatcher. Birdsong is a symphony: intricate, beautiful, and he can pick out all the players. He can tell me what he is looking at from the shape of a wing in flight. It's a knowledge not instantly graspable, but apparently infinitely rewarding. In lockdown, urban Americans became birdwatchers, too, in their thousands. For some at least, it was a form of self-care: a slow-paced hobby for a slower-paced existence. But it served also as insight, into another form of life, another way of being.[26]

That doesn't surprise Andrews. "If you're on your walk and you know about bird language then you can know that that's an alarm call, that one's started singing. It's gone beyond just naming. It's about how it's giving you information about their lives and how it's happening. This is a community of beings and you want to get to know them."

"I think it's an act of love," she tells me, "to be paying attention." For parents, surely, that's doubly true.

SELF-CARE (BUT NOT AS COMMODITY)

When Paul Hoggett and Rosemary Randall interviewed climate activists, they all described a similar experience. They had some kind of "epiphany," realizing the severity of the crisis. (For many parent climate activists, this was the 2018 IPCC report.) They became totally immersed: reading, thinking, talking, acting. Then came personal crises: burnout and overwhelm, being pounded by conflicting emotions. Finally, they made some adjustments, worked on their sense of agency, managed to balance climate action with living their lives, avoided constant preoccupation with hard facts, and achieved a sort of resolution.[27]

Getting through this process means setting boundaries, learning when to back off or temporarily suppress emotions, and when to reengage. This, says Andrews, takes mindfulness. It also requires self-care, but that *doesn't* mean the kind of commoditized self-care that is often taken for granted: the expensive products and five-star hotels that advertisers tell us we need.

I *like* spas and nice restaurants; I'm not going to pretend I don't. But, for me, self-care is seeking professional help if I'm sick, in pain, or struggling with anxiety. It's talking to close friends or family. It's twenty minutes walking or running through park or woods, snatched from the middle of the working day. It's a rare weekend entirely on my own. It's a bubble bath, a cup of coffee in the garden, cuddling my girls, or watching TV with the cat on my knee.

Other parent activists talk of mountain hikes and local walks, of the people they love, and of each other. "I need people I can be real with," says Shugarman. "Emotional health is a challenge," says Fuyani, "but I am very blessed to be surrounded by incredible women who support me, who truly care and love me. I have this huge box of things I know can help. I meditate. I do yoga, I hike, I swim, I try to be around positive vibrant people. I dance."

Connecting with nature, and other people. In other words, what we should be doing anyway, if we're serious about finding a psychological way forward.

We face four psychological tasks.

We must address hard emotions. It takes time, working through the shame, fear, grief, and rage that go with acknowledging the state of our world. It takes friends, family, and fellow activists. It might take professional help. But these emotions, constantly suppressed, keep us in a spiral of denial and mounting anxiety. If we do this work, we might just come out the other side: to determined hope.

We must overcome short-term thinking. As parents, we have a powerful motivation to care for the future: we love our kids. But for "deep care," we need longer-term thinking. We can practice casting our minds across centuries and millennia, to see ourselves as future historians and geologists will see us. In privileged, polluting societies, we might also try to see ourselves as our global contemporaries do.

The nonhuman world is no exploitable commodity; we need it, and we are a part of it. To reconnect, individually, we must abandon our props and teach ourselves to see, listen, and notice. Our children, if they are young enough, already know.

Finally, we must look after ourselves. That doesn't mean consumerist, packaged indulgence. It means learning to step back, when needed, seeking people to be yourself with, and time to be yourself in. It means loving your children and enjoying them, in the moment. For many, it means spending time outside: breathing, running, watching, lying on the sand. Finding balance.

CONCLUSION

May 12, 2022. I sit at my desk in a university building, looking out at the round towers and spiky roofs of what was once a hospital. Inspired by *Dear Tomorrow*, I've decided to write a letter to my daughters in 2050. But when I pick up my pen, I can't do it.

My brain is too crowded with images of a mid-century that I may not live to see. Of fear. Of ever-extended lockdowns. Of infections uncontrolled and uncontrollable. Of the worsening oppression of millions of human beings. Of miles of burning forest, catching up everything in its wake. Of homes disappearing under floods.

And images of hope. Of societies living with the environment rather than against it. Of recognition and justice. Of adults and children who find self-worth outside of material possessions. Of bikes and trains and children running safely through green cities, breathing clean air.

The year 2050 is a turning point. It's when greenhouse gas emissions must hit net zero, to avoid devastating climate change. It is also the year my daughters will turn thirty-seven and thirty-five. They will have to decide soon enough, if they haven't already, whether to bring new people into the world. I'm a writer, but when I try to speak to the people I love most, in that uncertain moment, the words won't come.

Instead, I read back over what I've written here.

Staring into collective disaster, it's easy to feel helpless. We grip our toddlers' hands, we call our teenagers back, as the tides creep in. We feel sad, frustrated. We feel shame, perhaps passionate anger. In the face of emergency on the grand scale, it can seem as though we have only two options: cling with eyes closed to the here and now, or fatalistically prepare our own children for a stark future. I hope I've shown why this is wrong.

If I owe my kids anything at all, beyond keeping them alive *as* children, I owe it to them protect their future. I've tried to show that here. I've tried to show, too, that this doesn't pull against the basic moral duties I have as a human being. In fact, it goes hand in hand with fulfilling them. My children need a world in which they can live and flourish, *and* in which their children and grandchildren can thrive. But they are (or will be) global citizens too. A just world is a better world for them. So is a living, thriving natural one. I cannot build this alone but, as parents and citizens, we can do almost anything together. And we should.

As part of this work, I must help my girls understand the challenges that face them, elaborated throughout this book: climate change; antibiotic resistance; pandemics; institutionalized injustice. I must raise my children to be motivated more by morality, less by materialism, and I must do all this while building their ability to live and think for themselves. I must adjust what we do, day by day, as a family. Most of all, I should be part of challenging institutions, holding politicians to account, and changing the way we *all* live. So, I think, should you. We owe it to our own children, as well as to everyone else.

The one-sentence summary? To be a good parent, in this threatened world, I must be an agent for change.

It's a long-haul project, forging a better world. It means carving time out of already packed days, overhauling things we take for granted. It means accepting hard truths and difficult emotions, holding both anxiety and hope in our hearts. It means facing up to the many and real moral dilemmas that I have not been able to philosophize away. I've tried to find the philosophical and psychological tools to do all this, and to lay them clearly, so you can do the same.

I recall all this, on this beautiful Edinburgh day. I remember something else, too. None of this means being unable to enjoy our own lives, or the incredible joy our children give us. We can even enhance them both, along the way. And this is an opportunity, as well as a crisis. "You are the only source of hope to your children and grandchildren," Ugandan activist Herbert Murungi told me nine months ago, from half a world away. I hear his words: "You must choose differently." I pick up my pen.

"I love you," I write to my daughters in twenty-eight years' time. "I promise I'll do my best."

Acknowledgments

The idea for this book emerged, mid-lockdown, in a Zoom conversation with my wonderful agent, Jaime Marshall. I'm incredibly grateful for his unceasing enthusiasm, inspiration, and expertise, and that of my fantastic MIT editor, Beth Clevenger. I could not have done this without them. Thank you, too, to the rest of the amazing team at the MIT Press, including Kathleen Caruso, Julia Collins, Elisabeth Graham, Jessica Pellien, and Anthony Zannino.

I've written this as academic, journalist, and parent: a juggling act as rewarding as it has been difficult. One highlight was interviewing some incredible scholars, parents, and activists. Huge thanks go to Nadine Andrews, Benji Backer, Leanne Brummell, Otto Cars, Susan Clayton, Xoli Fuyani, Sophie Harman, Lisa Howard, Cathrine Jansson-Boyd, Brian Kateman, Maya Mailer, Herbert Murungi, Sarah Myhre, Meredith Niles, and Harriet Shugarman. I have also drawn liberally on the written wisdom and experience of philosophers, scientists, sociologists, psychologists, activists, and, of course, parents. I gratefully acknowledge this.

Four brilliant thinkers and activists were kind enough to read part or all of a draft of the book, for a workshop at the University of Edinburgh, organized and chaired by my lovely colleagues Mihaela Mihai and Philip Cook. Thank you to Tim Fowler, Marlies Kustatscher, Harriet Shugarman (again!), and Adam Swift. I've been an academic and a parent long enough to know how big an ask it is to find the time for this in a busy teaching term, and how much better the book is as a result. Two further anonymous reviewers read the next draft for the MIT Press, and I am deeply grateful for their comments.

Several of the kind parents and grandparents in my life made time to read some or all of a later draft, even though I only gave them a week to do it! Thank you to Tom Baird, Jenny Bos, Ros Claase, Harry Cripps, Rosie Cripps, Vivien Cripps, Sarah Jones, Alan Saunders, and Kirstie Skinner, for an amazing combination of editorial insights and much-needed reassurance.

Thank you to my wonderful mentor, Fiona Mackay. To Maya Mailer and everyone on the Our Kids' Climate calls, for inspiring and welcoming me. To the friends, family, and colleagues who have helped and encouraged me while I've been writing, especially Mum, Dad, Sarah, Kate, and David. Thanks also to Dave, for lending me the perfect bolt hole for a week of frantic writing. To Helen and Alan, who put me up (and put up with me!) on every London trip. To Brian and Marly, who de-stress me. To Tom, my partner in these hopes and hard choices, who has done more than his share of actual parenting in the past eighteen months, so I can write about it. To all my nephews and nieces, actual and implicit, who deserve a beautiful world. Most of all, to my inspirational daughters, for whom I want everything, and who make this all worthwhile.

Notes

INTRODUCTION

1. I borrow the notion of "commonsense morality" from Samuel Scheffler, *Boundaries and Allegiances: Problems of Justice and Responsibility in Liberal Thought* (Oxford: Oxford University Press, 2001).

2. On the war, see Derek Saul, "Over 3,000 Civilians Killed in Ukraine since Russia Invaded, U.N. Says," *Forbes,* May 2, 2022, https://www.forbes.com /sites/dereksaul/2022/05/02/over-3000-civilians-killed-in-ukraine-since -russia-invaded-un-says/; "How Many Ukrainians Have Fled Their Homes and Where Have They Gone?," BBC News, May 6, 2022. Shute's novel deals with the aftermath of nuclear war, the spread of radiation, and the ways the characters, including a couple with a baby daughter, end their lives. Nevil Shute, *On the Beach* (London: Heinemann, 1957).

3. According to some commentators, the Ukraine invasion was also predictable. However, it resulted largely from the personality of one man. That separates it from these other global emergencies (the impact of Trump and Bolsonaro on climate policy notwithstanding). See, for example, Keith Gessen, "Was It Inevitable? A Short History of Russia's War on Ukraine," *The Guardian*, March 11, 2022; Ross Douthat, "They Predicted the Ukraine War. But Did They Still Get It Wrong?," *New York Times*, March 9, 2022.

4. The IPCC report is *Climate Change 2021: The Physical Science Basis. Contribution of Working Group 1 to the Sixth Assessment Report of the Intergovernmental Panel on Climate Change* (Cambridge: Cambridge University Press, 2021). (The figures on heat extremes that follow are from this report.)

More than a decade ago, Peter Doran and Maggie Kendall Zimmerman surveyed 3,146 scientists on climate change; 90 percent agreed that it was real and 82 percent that human activity had been significant in causing it. Among climate scientists, these figures were 96.2 percent and 97.4 percent; see Doran and Zimmerman, "Examining the Scientific Consensus on Climate Change," *Eos, Transactions American Geophysical Union* 90, no. 3 (2009): 22–23. Fore is quoted in Damian Carrington, "A Billion Children at 'Extreme Risk' from Climate Impacts," *The Guardian*, August 20, 2021.

5. Unless specified otherwise, the quotes from Shugarman are from a Zoom interview with the author on August 31, 2021.

6. The call for 45 percent cuts by 2030 is from IPCC, *Global Warming of 1.5°C: An IPCC Special Report on the Impacts of Global Warming of 1.5°C above Pre-Industrial Levels and Related Global Greenhouse Gas Emissions Pathways, in the Context of Strengthening the Global Response to the Threat of Climate Change, Sustainable Development, and Efforts to Eradicate Poverty* (Geneva, Switzerland: World Meteorological Organization, 2018). UNICEF's report is *The Climate Crisis Is a Child Rights Crisis: Introducing the Children's Climate Risk Index* (New York: United Nations Children's Fund, UNICEF, 2021). The 250,000 figure is from the World Health Organization, "Climate Change and Health," fact sheet, 2021, https://www.who.int/news-room/fact-sheets/detail/climate-change-and-health.

7. The study is Elizabeth Marks et al., "Young People's Voices on Climate Anxiety, Government Betrayal and Moral Injury: A Global Phenomenon," *The Lancet* (preprint 2021). Fossil fuel air pollution killed more than eight million people in 2018. See Karn Vohra et al., "Global Mortality from Outdoor Fine Particle Pollution Generated by Fossil Fuel Combustion: Results from Geos-Chem," *Environmental Research* 195 (2021), https://doi.org/10.1016/j.envres.2021.110754. In the two hundred years since the Industrial Revolution, oceans have become 30 percent more acidic. See National Oceanic and Atmospheric Administration, "Ocean Acidification," https://www.noaa.gov/education/resource-collections/ocean-coasts/ocean-acidification, last updated April 1, 2020.

8. The health workers' quotes are from Barkha Mathur, "National Doctors' Day 2021: 'Like Going to War': A Young Doctor Describes Working Covid Ward during the Second Wave," July 1, 2021, NDTV, https://swachhindia.ndtv.com/national-doctors-day-2021-like-going-to-war-a-young-doctor

-describes-working-in-covid-ward-during-the-second-wave-60722/; Shaun Lintern, "'Like Going to War': Life and Death on a Covid Intensive Care Ward,'" *The Independent*, February 5, 2021. The repeated simile is telling. Figures on COVID-19 deaths are taken from World Health Organization, "14.9 Million Deaths Associated with the COVID-19 Pandemic in 2020 and 2021," news release, May 5, 2022, https://www.who.int/news/item/05 -05-2022-14.9-million-excess-deaths-were-associated-with-the-covid-19 -pandemic-in-2020-and-2021.

9. Quotes from Harman are all from a Zoom interview with the author on September 30, 2021. The quote from Brilliant is from "Interview: Larry Brilliant. How Society Can Overcome Covid-19," *The Economist*, April 4, 2020.

10. Schools were closed from seven to nineteen weeks across OECD and partner countries between February 17 and June 30, 2020. See Andreas Schleicher, *The Impact of COVID-19 on Education: Insights from "Education at a Glance 2020"* (Paris: OECD, 2020). The statistics are from Mind, *The Mental Health Emergency: How Has the Coronavirus Pandemic Impacted Our Mental Health?* (London: Mind, 2020), and Cathy Creswell et al., "Young People's Mental Health during the COVID-19 Pandemic," *The Lancet Child & Adolescent Health* 5, no. 8 (2021): 535–537.

11. On the Nevada death, see Lei Chen, Randall Todd, Julia Kiehlbauch, Maroya Walters, and Alexander Kallen, "Notes from the Field: Pan-Resistant New Delhi Metallo-Beta-Lactamase-Producing *Klebsiella pneumoniae*—Washoe County, Nevada, 2016," *Morbidity and Mortality Weekly Report* (Atlanta: Centers for Disease Control and Prevention, 2017).

12. The sources are CDC, *Antibiotic Resistance Threats in the United States 2019* (Atlanta: US Department of Health and Human Services, CDC, 2019); Alessandro Cassini et al., "Attributable Deaths and Disability-Adjusted Life-Years Caused by Infections with Antibiotic-Resistant Bacteria in the EU and the European Economic Area in 2015: A Population-Level Modelling Analysis," Infectious Diseases, *The Lancet* 19, no. 1 (2019): 56–66; Jim O'Neill, *Tackling Drug-Resistant Infections Globally: Final Report and Recommendations* (London: Wellcome Trust and HM Government, 2016); Nesta, "Superbugs Put Future Viability of Cancer Treatment into Question," press release, February 18, 2020, https://www.nesta.org.uk/press-release/super bugs-put-future-viability-cancer-treatment-question-new-study/.

13. The World Health Organization quote is from "Antimicrobial Resistance," fact sheet, November 17, 2021, https://www.who.int/news-room/fact -sheets/detail/antimicrobial-resistance. Quotes from Otto Cars are all taken from a Zoom interview with the author, September 22, 2021. In 2019, only thirty-two antibiotics were in development that would target priority pathogens. Only six could be classed as innovative. A 2022 joint study by WHO and the European Centre for Disease Control and Prevention confirmed that antibiotic resistance remains a major public health concern and highlighted major drivers: *Antimicrobial Resistance Surveillance in Europe 2022: 2020 Data* (Geneva: WHO and ECDCP, 2022).

14. Mary Annaïse Heglar, "Climate Change Isn't the First Existential Threat," *Medium*, February 18, 2019, https://zora.medium.com/sorry-yall-but-cli mate-change-ain-t-the-first-existential-threat-b3c999267aa0.

15. The statistics on violence against women are taken from World Health Organization, *Violence against Women Prevalence Estimates, 2018* (Geneva: WHO, 2021). Those on US convictions are from the Rape, Abuse & Incest National Network (RAINN), "The Criminal Justice Statistics," https:// www.rainn.org/statistics/criminal-justice-system. It is hugely difficult to be accurate on sexual assault figures when the majority are not even reported. However, RAINN uses the National Crime Victimization Survey as a primary data source. The prevalence of sexual harassment in UK schools was confirmed by the schools' regulatory body; see Ofsted, *Review of Sexual Abuse in Schools and Colleges* (London: UK Government, 2021). On *Roe v. Wade*, see, for example, Josh Gerstein, Alice Miranda Ollstein and Quint Forgey, "Supreme Court Gives States Green Light to Ban Abortion, Overturning Roe," *Politico*, June 24, 2022. The white male student was Brock Turner, the unconscious woman he assaulted was Chanel Miller; see Lynn Neary, "Victim of Brock Turner Sexual Assault Reveals Her Identity," *NPR*, May 21, 2019.

16. For the news coverage of these murders, see, for example, Biba Adams, "Texas Man Accused of Killing Wife, Firing at Children after She Made Plans to Leave," *Yahoo! News*, August 4, 2021; Vikram Dodd and Haroon Siddique, "Sarah Everard Murder: Wayne Couzens Given Whole-Life Sentence," *The Guardian*, September 30, 2021. The statistics are taken from Sally Martinelli, "Nearly 2,000 Women Murdered by Men in One Year, New Violence Policy Center Study Finds," Violence Policy Center, news

release, September 23, 2020, https://vpc.org/press/nearly-2000-women
-murdered-by-men-in-one-year-new-violence-policy-center-study-finds/.

17. Young was interviewed by Sam Sanders: "A Black Mother Reflects on Giving
    Her 3 Sons 'the Talk' . . . Again and Again," *NPR*, June 28, 2020.

18. The statistics are from the Washington Post, "1,057 People Have Been
    Shot and Killed by Police in the Past Year," *Washington Post*, https://www
    .washingtonpost.com/graphics/investigations/police-shootings-database/
    (accessed August 17, 2022); Maria Trent et al., "The Impact of Racism on
    Child and Adolescent Health," *Pediatrics* 144, no. 2 (2019), https://doi
    .org/10.1542/peds.2019-1765; Emma Garcia, *Schools Are Still Segregated,
    and Black Children Are Paying a Price* (Washington, DC: Economic Policy
    Institute, 2020). Black students in high-poverty schools score an average of
    twenty points lower on standardized scores than their contemporaries in
    low-poverty, mostly white schools.

19. As a concept, intersectionality illustrates how people's different politics and
    social identities combine, silencing them, magnifying discrimination, or (on
    the flipside) increasing their privilege. See Kimberlé Crenshaw, "Demar-
    ginalizing the Intersection of Race and Sex: A Black Feminist Critique of
    Antidiscrimination Doctrine, Feminist Theory and Antiracist Politics," *Uni-
    versity of Chicago Legal Forum* 140 (1989): 139–167. On hate crimes and
    LGBT kids, see Chaka L. Bachmann and Becca Gooch, "LGBT in Britain:
    Hate Crime and Discrimination" (London: Stonewall and YouGov, 2017).

20. The statistics on child poverty, labor, and malnutrition are from the World
    Bank Group and UNICEF, *Ending Extreme Poverty: A Focus on Children* (New
    York and Washington, DC: UNICEF, World Bank Group, 2016); World
    Health Organization, "Children: Improving Survival and Well-Being," Sep-
    tember 8, 2020, https://www.who.int/news-room/fact-sheets/detail/chil
    dren-reducing-mortality; UNICEF, *The State of the World's Children 2019.
    Children, Food and Nutrition: Growing Well in a Changing World* (New York:
    UNICEF, 2019); International Labour Office, *Marking Progress against
    Child Labour* (Geneva: ILO, 2013).

21. The United Nations highlights the disproportionate impact of climate
    change on women, including through domestic violence. See United
    Nations Framework Convention on Climate Change, "Differentiated
    Impacts of Climate Change on Women and Men; the Integration of

Gender Considerations in Climate Policies, Plans and Actions; and Progress in Enhancing Gender Balance in National Climate Delegations," June 12, 2019, https://unfccc.int/documents/196305. On the impact of home schooling, see Richard J. Petts, Daniel L. Carlson, and Joanna R. Pepin, "A Gendered Pandemic: Childcare, Homeschooling, and Parents' Employment During Covid-19," *Gender, Work & Organization* 28, no. S2 (2021): 515–534; B. Xue and A. McMunn, "Gender Differences in Unpaid Care Work and Psychological Distress in the UK Covid-19 Lockdown," *PLoS One* 16, no. 3 (2021), https://doi.org/10.1371/journal.pone.0247959.

22. Of the estimated 1.27 million deaths attributable to antimicrobial resistance in 2019, the highest rate (27.3 per 100,000) was in Sub-Saharan Africa. See Christopher J. L. Murray et al., "Global Burden of Bacterial Antimicrobial Resistance in 2019: A Systematic Analysis," *The Lancet* 399, no. 10325 (2022): 629–655. However, more people in low- and middle-income countries die from lack of access to antibiotics than from antibiotic resistance. See Aditi Sriram et al., *State of the World's Antibiotics 2021: A Global Analysis of Antimicrobial Resistance and Its Drivers* (Washington, DC: Center for Disease Dynamics, Economics and Policy, 2021). The IPCC warnings are from its "Summary for Policymakers," in *Climate Change 2014: Mitigation of Climate Change. Contribution of Working Group III to the Fifth Assessment Report of the Intergovernmental Panel on Climate Change*, ed. O. Edenhofer et al. (Cambridge and New York: Cambridge University Press, 2014), 1–48. I discuss the intersectional impact of climate change and its roots in slavery and colonialism in chapter 2 of my book *What Climate Justice Means and Why We Should Care* (London: Bloomsbury, 2022), 49–71.

The statistics on vaccination rates are from Our World in Data, "Coronavirus (Covid-19) Vaccinations," Oxford Martin School and Global Change Data Lab, https://ourworldindata.org/covid-vaccinations (accessed August 17, 2022). Those on climate change, COVID-19, and poverty are from UNICEF Data, "Impact of COVID-19 on Multidimensional Child Poverty" (September 2020), https://data.unicef.org/resources/impact-of-covid-19-on-multidimensional-child-poverty/; Julie Rozenberg and Stephane Hallegatte, "Poor People on the Front Line: The Impacts of Climate Change on Poverty in 2030," in *Climate Justice: Integrating Economics and Philosophy*, ed. Henry Shue and Ravi Kanbur (Oxford: Oxford University Press, 2018), 24–42.

23. The statistics on the impact of COVID-19 on the vulnerable are from "Risk for COVID-19 Infection, Hospitalization, and Death by Race/Ethnicity," Centers for Disease Control and Prevention, https://www.cdc.gov /coronavirus/2019-ncov/covid-data/investigations-discovery/hospitaliza tion-death-by-race-ethnicity.html (accessed August 17, 2022); Poor People's Campaign, *A Poor People's Pandemic Report: Mapping the Intersection of Poverty, Race and Covid-19* (Poor People's Campaign, Kairos Center, Repairers of the Breach, SDSN, Howard University, 2022). On the rising demand for UK food banks, see Miranda Bryant, "'We're Terrified at What We're Seeing': Food Banks Tell of Soaring Demand," *The Guardian*, February 6, 2022.

CHAPTER 1

1. For Kant's core argument, see Immanuel Kant, *Groundwork for the Metaphysics of Morals* (Oxford: Oxford University Press, [1785] 2019).

2. Aristotle's virtue ethics is from "Nicomachean Ethics," in *The Complete Works of Aristotle*, ed. Jonathan Barnes (Princeton: Princeton University Press, 1984).

3. The thought experiment is based on Robert Nozick's hypothetical "experience machine" in *The Examined Life: Philosophical Meditations* (New York: Simon and Schuster, 1989), 104–105. For Hurka's argument, see *The Best Things in Life: A Guide to What Really Matters* (Oxford and New York: Oxford University Press, 2011), 67–73.

4. This point is made by philosophers Jennifer Wilson Mulnix and M. J. Mulnix in their engaging book *Happy Lives, Good Lives: A Philosophical Examination* (Ontario: Broadview Press, 2015), 175–184.

5. On basic needs, see, for example, Frances Stewart, "Basic Needs, Capabilities, and Human Development," in *In Pursuit of the Quality of Life*, ed. Avner Offer (New York: Oxford University Press, 1996), 46–65. The quote is from Amartya Sen, *Development as Freedom* (Oxford: Oxford University Press, 1999), 74. I draw particularly on Nussbaum's capabilities approach, which Brighouse and Swift adapt for children: Martha Nussbaum, *Women and Human Development: The Capabilities Approach* (Cambridge: Cambridge University Press, 2000), 78–80; Harry Brighouse and Adam Swift, *Family Values: The Ethics of Parent-Child Relationships* (Princeton: Princeton University Press, 2014), 41–43. Nussbaum developed her list using a combination

of Aristotelian philosophy and engagement with women living in poverty in India, and thinks people with different religions or cultures could agree on it.

6. For Mill's no-harm principle, see "On Liberty" in *John Stuart Mill: On Liberty and Other Essays*, ed. John Gray (Oxford: Oxford University Press, 1859), 14. Mill defends his principle indirectly. He argues that states can only justify interfering with individuals' liberty in order to prevent harm to others. For Kant's categorical imperative, see *Groundwork for the Metaphysics of Morals*. Rosalind Hursthouse argues that virtue ethics is just as helpful in guiding individual behavior as utilitarianism or Kantian ethics: *On Virtue Ethics* (Oxford: Oxford University Press, 1999), 35–39.

7. Singer's paper is "Famine, Affluence, and Morality," *Philosophy and Public Affairs* 72, no. 1 (1972): 229–243. Some philosophers, like many people, think we owe less to our non-compatriots than to those who share our states. But even they rarely deny these core duties of humanity. For an example of a philosopher who rejects the idea of global, institutionalized *equality* but accepts duties to aid the needy, see Thomas Nagel, "The Problem of Global Justice," *Philosophy and Public Affairs* 33, no. 2 (2005): 113–147.

8. The Hursthouse quote is from *On Virtue Ethics*, 36.

9. This last claim is *not*, of course, that mums or dads are "bad parents" if they don't feel an instant, overwhelming bond with their baby. A third of UK mothers find it difficult to build that initial bond; see National Childbirth Trust, "Difficulties with Baby Bonding Affect a Third of UK Mums" (London: National Childbirth Trust, 2016). It is, rather, that we shouldn't be too rose-tinted about parents in general. According to the American Society for the Prevention of Cruelty to Children, 1,750 children died from abuse and neglect in 2020; more than 90 percent of reports of abuse were by one or both parents: American SPCC, "Child Maltreatment Statistics," https://americanspcc.org/child-abuse-statistics/ (accessed August 17, 2022).

10. The two ways of explaining parents' special duties are known as "causal" and "intentional" (sometimes also called "voluntarist") accounts. See Jeffrey Blustein, "Procreation and Parental Responsibility," *Journal of Social Philosophy* 28, no. 2 (1997): 79–86; Onora O'Neill, "Begetting, Parenting, and Rearing," in *Having Children: Philosophical and Legal Reflections on Parenthood*, ed. Onora O'Neill and William Ruddick (New York: Oxford

University Press, 1979), 25–38; Elizabeth Blake, "Willing Parents: A Voluntarist Account of Parental Role Obligations," in *Procreation and Parenthood: The Ethics of Bearing and Rearing Children*, ed. David Archard and David Benetar (Oxford: Oxford University Press, 2010), 151–177.

11. David Archard discusses discharging causally acquired parental responsibilities by passing them onto someone else in "The Obligations and Responsibilities of Parenthood," in *Procreation and Parenthood: The Ethics of Bearing and Rearing Children*, ed. David Archard and David Benetar (Oxford: Oxford University Press, 2010), 103–127.

12. For Brighouse and Swift's argument, see *Family Values*.

13. The practitioners' view is from Joanne Kellett and Joanna Apps, eds., *Assessments of Parenting and Parenting Support Need: A Study of Four Professional Groups* (York: Joseph Rowntree Foundation, 2009).

14. For more on play, see Kathleen Glascott Burriss and Ling-Ling Tsao, "Review of Research: How Much Do We Know About the Importance of Play in Child Development?," *Childhood Education* 78, no. 4 (2002): 231.

15. As Brighouse and Swift put it: "The parent's fiduciary duties are to guarantee the child's immediate well-being, including . . . the intrinsic goods of childhood, and to oversee her cognitive, emotional, physical, and moral development" (*Family Values*, 53). Care ethicist Sara Ruddick describes "maternal thinking" as arising from children's need for both preservation and growth. See "Maternal Thinking," *Feminist Studies* 6, no. 2 (1980): 342–367.

16. In defending the family and calling for more than impersonal care, Brighouse and Swift draw on both attachment theory and neuroscience (*Family Values*, 44–47).

17. Leading work in care ethics includes S. Ruddick, "Maternal Thinking"; Eva Feder Kittay, *Love's Labor: Essays on Women, Equality and Dependency*, 2nd ed. (Abingdon, UK: Taylor & Francis, 2020); Nel Noddings, *Caring, a Feminine Approach to Ethics & Moral Education* (Berkeley: University of California Press, 1984). This is not limited to personal matters: it provides an alternative ethical approach to broader political and global relations that could overlap with the notion of a "good global citizen" filled out in this book. See, for example, Virginia Held, *The Ethics of Care: Personal, Political, and Global* (New York: Oxford University Press, 2006); Nancy Folbre, *For*

Love and Money: Care Provision in the United States (New York: Russell Sage Foundation, 2012).

18. The details of the Black Friday tragedy are taken from Lukas I. Alpert, "Victim's Life a Struggle," *New York Post*, November 29, 2008; "Wal-Mart Worker Killed in Black Friday Shopping Stampede," *The Guardian*, November 29, 2008.

19. I talk more about collectivizing the no-harm principle in my book *Climate Change and the Moral Agent: Individual Duties in an Interdependent World* (Oxford: Oxford University Press, 2013).

20. On "new harms," see Judith Lichtenberg, *Distant Strangers: Ethics, Psychology, and Global Poverty* (New York: Cambridge University Press, 2014), 73.

21. For Gardiner's argument, see "Is No-One Responsible for Global Environmental Tragedy? Climate Change as a Challenge to Our Ethical Concepts," in *The Ethics of Global Climate Change*, ed. Denis G. Arnold (Cambridge: Cambridge University Press, 2011).

22. The quote is from May, *Sharing Responsibility* (Chicago and London: University of Chicago Press, 1992), 121.

23. For the Durdle Door news story, see Emily Foote, "Human Chain Rescues Man Struggling in Sea at Durdle Door—Video," *The Guardian*, August 23, 2020. Often, of course, such ad hoc rescues are impossible or unreasonably dangerous without an adequate safety kit.

24. May, *Sharing Responsibility*, 105–124. Other philosophers have also defended duties to organize to save lives, including care ethicist Virginia Held, "Can a Random Collection of Individuals Be Morally Responsible?," *Journal of Philosophy* 67, no. 14 (1970): 471–481; utilitarian Robert Goodin, *Protecting the Vulnerable: A Reanalysis of Our Social Responsibilities* (Chicago and London: University of Chicago Press, 1985); and cosmopolitan global justice scholar Henry Shue, *Basic Rights: Subsistence, Affluence, and U.S. Foreign Policy* (Princeton: Princeton University Press, 1980). I collectivize the principle of beneficence, drawing on these sources, in my book *Climate Change and the Moral Agent*.

25. The news story is Geert De Clercq and Ingrid Melander, "Twenty-Seven Migrants Perish Trying to Cross Channel to Britain," *Reuters*, November 24, 2021.

26. See Shue, *Basic Rights*. For a detailed exploration of the duties of affluent individuals in the face of global poverty, see Christian Barry and Gerhard Øverland, *Responding to Global Poverty: Harm, Responsibility, and Agency* (Cambridge: Cambridge University Press, 2016).

27. This idea of global citizenship owes a lot to cosmopolitan thinkers, in particular Henry Shue's idea that human communities have a responsibility to design institutions that protect people's basic rights; see Shue, *Basic Rights*; see also Charles R. Beitz and Robert E. Goodin, *Global Basic Rights* (Oxford: Oxford University Press, 2011). On global/environmental citizenship, see, for example, Andrew Dobson, *Citizenship and the Environment* (Oxford: Oxford University Press, 2003).

CHAPTER 2

1. These tragic stories are told in Jessica Elgot, "Family of Syrian Boy Washed up on Beach Were Trying to Reach Canada," *The Guardian,* September 3, 2015; Declan Walsh, "Yemen Girl Who Turned World's Eyes to Famine Is Dead," *New York Times*, November 1, 2018; "The Tragedy of Saudi Arabia's War," *New York Times*, October 26, 2018. On the accountability of the United States, see, for example, Andrea Prasow, "US War Crimes in Yemen: Stop Looking the Other Way," Human Rights Watch, September 21, 2020, https://www.hrw.org/news/2020/09/21/us-war-crimes-yemen-stop-looking-other-way.

2. The survey on the cost of toys was by the Toy Industry Association; see SWNS Digital, "Average Child Gets $6,500 Worth of Toys in Their Lifetime," *SWNS Digital*, September 6, 2021. The data on kids' first birthdays is from BabyCenter, "How Much Did You Spend on Your Child's First Birthday Party?," BabyCenter, https://www.babycenter.com/4_how-much-did-you-spend-on-your-childs-first-birthday-party_1490017.bc. A 2017 study found that the average UK parents spent £500 on their child's birthday (party and presents): "The Average British Child's Birthday Party Costs over £320," Vouchercloud, https://www.vouchercloud.com/resources/cost-of-childrens-parties. On the trend for extortionate toddler birthday parties, see Alisa Wolfson, "Some Toddler Birthday Parties Now Cost as Much as Weddings. Parents Are Dropping Hundreds of Thousands—Because Their Kids Only Turn 3 Once," *Insider*, January 29, 2020.

3. The comments are from Emma Gill, "Mum Shamed on Facebook after Sharing Daughter's Christmas Present Pile," *Manchester Evening News*, December 19, 2020. The mother in the story stressed that her daughter "is not a brat" and does community and volunteer work. The unacceptability of the public "shaming" of this one woman in particular is, of course, a separate question from the one discussed here.

4. I'm not suggesting that nonparents all do enough to tackle global suffering, and parents are the exception. Far from it. There's some evidence that parents give less to charity or community projects than those without kids, but it's equivocal. Analysts found that almost a quarter of childless British people over age fifty and 16 percent in the Netherlands had written a charitable will, compared to 8 percent and 4 percent of those with children; see Meg Abdy, "Lessons from Abroad," Legacy Foresight (2016), https://www.legacy foresight.co.uk/viewpoint/3183146-lessons_abroad/. But Michael Hurd, director of the RAND Centre for the Study of Aging, found no consistent difference in charitable giving between parents and nonparents in couples, though single childless people tended to give slightly more to charity than single people with children; see "Inter-Vivos Giving by Older People in the United States: Who Received Financial Gifts from the Childless?," *Ageing and Society* 29, no. 8 (2009): 1207–1225. Instead, my point is this. We tend to think there's something special about parenting that makes this general failure permissible, even commendable, *for us*.

5. The quote is from "Moral Saints," *Journal of Philosophy* 79, no. 8 (1982): 419.

6. Judith Lichtenberg makes the point about being entitled to space to live one's own life (*Distant Strangers*). As Garrett Cullity puts it, there are "impartial" reasons to give these partiality grounding relationships some moral protection; see Cullity, *The Moral Demands of Affluence* (Oxford: Oxford University Press, 2004).

7. Singer also defends a much more demanding principle of beneficence than the one in the last chapter. He thinks I should sacrifice myself to help someone else up to the point that I would give up something of "comparable moral importance" ("Famine, Affluence, and Morality," 231). However, he is an act utilitarian. They are unusual in making no difference between costs and benefits to us, or those we love, and changes to overall welfare. Obviously, literature and art can also play an important role in highlighting and, ultimately, addressing injustice (Toni Morrison's work being a prime example).

8. Unger's book is *Living High and Letting Die* (New York: Oxford University Press, 1996).

9. This idea of ring-fencing comes from the philosopher Garrett Cullity (*The Moral Demands of Affluence.*) His is not the only approach. Liam Murphy thinks that no one can be expected to do more to help others than they would have to if everyone else was doing their part; see "The Demands of Beneficence," *Philosophy and Public Affairs* 22, no. 4 (1993): 267–292. I think that's a mistake. Fairness across duty bearers matters, but it's trumped by protecting basic human rights. See Dominic Roser and Sabine Hohl, "Stepping in for the Polluters? Climate Justice under Partial Compliance," *Analyse and Kritik* 33, no. 2 (2011): 477–500; see also my book *Climate Change and the Moral Agent*, 157–159. Or, as another philosopher puts it, how *much* we owe our fellow humans doesn't change depending on what other people do. What changes is how we're *actually* required to fulfill that duty; see Anja Karnein, "Putting Fairness in Its Place: Why There Is a Duty to Take Up the Slack," *Journal of Philosophy* 111, no. 11 (2014): 593–607.

10. The studies are Robert Epstein, "What Makes a Good Parent?," *Scientific American Mind* 21, no. 5 (2010); H. Breiner, M. Ford, and V. L. Gadsden, *Parenting Matters: Supporting Parents of Children Ages 0–8* (Washington, DC: National Academies Press, 2016), https://www.scientificamerican.com/article/what-makes-a-good-parent/.

11. For Clayton's argument, see *Justice and Legitimacy in Upbringing* (Oxford: Oxford University Press, 2006).

12. The Martinez novel is *The Oxford Murders*, trans. Sonia Soto (London: Little, Brown, 2005). Sorry for the spoiler.

13. Perry's quote is from her brilliant book *The Book You Wish Your Parents Had Read (and Your Children Will Be Glad That You Did)* (London: Penguin Life, 2019), 8.

14. For Brighouse and Swift's account of "relationship goods," see *Family Values*, 147.

15. The Brighouse and Swift quote is from *Family Values*, 122. For example, contrary to stereotype, kids needn't be disadvantaged by having two parents who work outside the home. See, for example, Kathleen McGinn, Mayra Ruiz Castro, and Elizabeth Lingo, "Learning from Mum: Cross-National Evidence Linking Maternal Employment and Adult Children's Outcomes,"

Work, Employment and Society 33 (2018): 374–400; R. G. Lucas-Thompson, W. A. Goldberg, and J. Prause, "Maternal Work Early in the Lives of Children and Its Distal Associations with Achievement and Behavior Problems: A Meta-Analysis," *Psychol Bull* 136, no. 6 (2010): 915–942.

16. The quotes from Jansson-Boyd are all from a Zoom interview with the author, September 6, 2021. According to one study, emotional well-being increases with income but only up to $75,000 a year: Daniel Kahneman and Angus Deaton, "High Income Improves Evaluation of Life but Not Emotional Well-Being," *Proceedings of the National Academy of Sciences* 107, no. 38 (2010), 16489–16493. However, a more recent study questions that upper limit: Matthew A. Killingsworth, "Experienced Well-Being Rises with Income, Even above $75,000 per Year," *Proceedings of the National Academy of Sciences* 118, no. 4 (2021): e2016976118, https://doi.org/10.1073/pnas.2016976118.

17. The Lichtenberg quote is from *Distant Strangers*, 136. She enlists both Karl Marx and Adam Smith (unusual bedfellows!) to make her point.

18. The "creepy" comment is from Brighouse and Swift, *Family Values*, 133. Swift tackles the contentious question of private education in *How Not to Be a Hypocrite: School Choice for the Morally Perplexed Parent* (London and New York: Routledge, 2003). I'll come back to this in chapter 8.

19. The statistics on global inequality are from Patricia Espinoza Revollo, Clare Coffey, Rowan Harvey, Max Lawson, Anam Parvez Butt, Kim Piaget, Diana Sarosi, Julie Thekkudan, *Time to Care: Unpaid and Underpaid Care Work and the Global Inequality Crisis* (Oxford: Oxfam International, 2020). In 2019, the world's 2,153 billionaires had more wealth among them than 4.6 billion people. The US child poverty level was a significant improvement on 2010 levels (22 percent), but both were pre-COVID-19 (United States Census Bureau data reproduced in Deja Thomas and Richard Fry, "Prior to COVID-19, Child Poverty Rates Had Reached Record Lows in U.S.," Pew Research Center, November 30, 2020, https://www.pewresearch.org/fact-tank/2020/11/30/prior-to-covid-19-child-poverty-rates-had-reached-record-lows-in-u-s/).

20. For Macleod's argument, see "Parental Responsibilities in an Unjust World," in Archard and Benetar, *Procreation and Parenthood*, 128–150.

21. The McGhee quotes are from her wonderful and (for me) eye-opening book *The Sum of Us: What Racism Costs Everyone and How We Can Prosper Together* (New York and London: Profile Books, 2021), 43.

22. The idea that children (and all of us) need a reliable opportunity to exercise central interests is drawn from Martha Nussbaum's capabilities model, introduced in chapter 1. See Nussbaum, *Women and Human Development*. Philosophers Jonathan Wolff and Avner De-Shalit discuss risk and flourishing in more depth in *Disadvantage* (Oxford: Oxford University Press, 2007).

23. The quote is from a previous version of the Mothers Rise Up website and was confirmed in an email from Gow to the author on May 16, 2022.

24. I talk in more philosophical detail about parents' climate duties in my article "Do Parents Have a Special Duty to Mitigate Climate Change?," *Politics, Philosophy & Economics* 16, no. 3 (2017): 308–325.

25. The Murdock quote is from Carl O'Brien, "Traffic Congestion Outside Schools Poses 'Incredibly Dangerous' Risks to Children," *Irish Times*, August 28, 2019. The statistics are from Harriet Edwards and Abigail Whitehouse, "The Toxic School Run" (London: UNICEF UK, 2018), 6.

26. Harriet Shugarman made this comment at an "Author Meets Critics" session on a draft of this book, run by CRITIQUE: Centre for Ethics and Critical Thought, University of Edinburgh, November 3, 2021.

27. Unless specified otherwise, the quotes from Howard are from a Zoom interview with the author on September 23, 2021. See also Lisa Howard, "When Global Problems Come Home: Engagement with Climate Change within the Intersecting Affective Spaces of Parenting and Activism," *Emotion, Space and Society* 44 (August 2022), https://doi.org/10.1016/j.emospa.2022.100894.

CHAPTER 3

1. The Hobbes quote is from *Leviathan* (Cambridge: Cambridge University Press, [1651] 1991), 89. It is his description of a state without government.

2. The Donne quote is from "Xvii. Meditation," in *John Donne: Devotions Upon Emergent Occasions Together with Death's Duel* (Ann Arbor: University of Michigan Press, [1623] 1959), 108.

3. Ray's concerns were based on the (predominantly white) responses to her book *A Field Guide to Climate Anxiety: How to Keep Your Cool on a Warming Planet* (Oakland: University of California Press, 2020). They are articulated in "Climate Anxiety Is an Overwhelmingly White Phenomenon," *Scientific American*, March 21, 2021. However, Nadine Andrews, climate psychologist, thinks climate anxiety can be a serious, clinical condition, experienced across cultures; from Zoom interview with the author, September 1, 2021. The *Lancet* study cited in chapter 1 included young people from Brazil, India, the Philippines, and Nigeria, as well as the UK, Finland, France, the United States, Australia, and Portugal. See Marks et al., "Young People's Voices on Climate Anxiety."

4. See Walsh, "Yemen Girl Who Turned World's Eyes." The other statistics are taken from Conway F. Saylor et al., "Media Exposure to September 11: Elementary School Students' Experiences and Posttraumatic Symptoms," *The American Behavioral Scientist* 46, no. 12 (2003): 1622–1642; Michał Bilewicz and Adrian Dominik Wojcik, "Visiting Auschwitz: Evidence of Secondary Traumatization among High School Students," *American Journal of Orthopsychiatry* 88, no. 3 (2018): 328–334.

5. The quote is from Donne, "Xvii. Meditation," 109.

6. The Harvey quote is from her book *Raising White Kids: Bringing up Children in a Racially Unjust America* (Nashville: Abingdon Press, 2017), 69–70.

7. Sarah Gotowiec and Elizabeth Cantor-Graae, a psychologist and a medical researcher, interviewed eight healthcare professionals; see Gotowiec and Cantor-Graae, "The Burden of Choice: A Qualitative Study of Healthcare Professionals' Reactions to Ethical Challenges in Humanitarian Crises," *Journal of International Humanitarian Action* 2, no. 2 (2017), https://doi.org/10.1186/s41018-017-0019-y.

8. This interpretation of "regret" is from Bernard Williams, *Moral Luck: Philosophical Papers* (Cambridge: Cambridge University Press, 1981); Carla Bagnoli, "Value in the Guise of Regret," *Philosophical Explanations* 3 (2000): 169–187. I discuss marring or "tragic" choices in chapter 7 of my book *Climate Change and the Moral Agent*, drawing on Thomas Nagel, *Equality and Partiality* (New York and Oxford: Oxford University Press, 1991), 1–10; Stephen M. Gardiner, "Is 'Arming the Future' with Geoengineering Really the Lesser Evil?," in *Climate Ethics: Essential Readings*, ed. Stephen

M. Gardiner et al. (Oxford and New York: Oxford University Press, 2010), 339–396; Stephen M. Gardiner, *A Perfect Moral Storm: The Ethical Tragedy of Climate Change* (Oxford and New York: Oxford University Press, 2011.

9. McGhee makes the point about draining the pool in *The Sum of Us*, 17–39.

10. The OECD report is John F. Helliwell et al., eds., *World Happiness Report 2020* (New York: Sustainable Development Solutions Network, 2020). The survey on welfare states and happiness is Patrick Flavin, Alexander Pacek, and Benjamin Radcliff, "Assessing the Impact of the Size and Scope of Government on Human Well-Being," *Social Forces* 92 (2014): 1241–1258. It is more ambiguous whether inequality itself undermines happiness all round. See, for example, Shigehiro Oishi, Selin Kesebir, and Ed Diener, "Income Inequality and Happiness," *Psychological Science* 22, no. 9 (2011): 1095–1100.

11. The Food and Agriculture Organization's warning is taken from FAO, *The FAO Action Plan on Antimicrobial Resistance 2021–2025* (Rome: FAO, 2021).

12. Adele tells her story in "Nightwatch," a blog post to *NWP Write Now*, June 22, 2020, https://writenow.nwp.org/night-watch-b7508a85ef7b. She describes herself as "on the right side of forty" when she found this out, in 2020, which makes this 1980s or 1990s America.

13. Atkar's story is taken from Simon Ingram, *A Gathering Storm: Climate Change Clouds the Future of Children in Bangladesh* (New York: UNICEF, 2019).

14. George Floyd was killed in Minneapolis on May 25, 2020, prompting a global uprising for the Black Lives Matter movement. See, for example, Ray Sanchez and Eric Levenson, "Derek Chauvin Sentenced to 22.5 Years in Death of George Floyd," *CNN*, June 25, 2021.

15. The cropped photo was taken at the Davos Economic Forum. The news agency later apologized, and said the crop was a mistake. See Kenya Evelyn, "'Like I Wasn't There': Climate Activist Vanessa Nakate on Being Erased from a Movement,'" *The Guardian*, January 29, 2020. On the solidarity strikes, see Fridays for Future Ukraine (@fff_ukraine), "Tomorrow, please support the activists and people of Ukraine by joining the worldwide #Fridays forfuture solidarity strikes," Twitter, March 2, 2022, 1:36 p.m., https://twitter.com/fff_ukraine/status/1499091398611677190.

16. For more on the demands for justice of youth climate movements, see "Our Demands," Fridays for Future, https://fridaysforfuture.org/what-we-do /our-demands/ (accessed November 10, 2022); "What We Believe: Sunrise's Principles," Sunrise Movement, https://www.sunrisemovement.org/ principles (accessed November 10, 2022).

17. See IPCC, *Climate Change 2021*, 14.

18. On environmental injustices to Indigenous communities, see, for example, Kyle Whyte, "Settler Colonialism, Ecology, and Environmental Injustice," *Environment and Society* 9, no. 1 (2018): 125–144. For a discussion of stewardship and intergenerational ethics, see Lawrence C. Becker, *Reciprocity* (London and New York: Routledge & Kegan Paul, 1986). For a basics-rights discussion of responsibilities to future generations, see Henry Shue, "Bequeathing Hazards: Security Rights and Property Rights of Future Humans," in *Global Environmental Economics: Equity and the Limits to Markets*, ed. Mohammed Dore and Timothy Mount (Oxford: Blackwell, 1999), 38–53.

19. The quotes from Sarah Myhre are taken from a Zoom interview with the author on September 1, 2021.

20. The "activist grannies" story is from BBC South: "Dorset County Hall Steps Up Security after 'Granny Invasion,'" BBC News, May 11, 2022.

21. The hunger strike was covered in Anoosh Chakelian, "'It's Hell': Meet the Starving Grandfather Outside Tory HQ Who Boris Johnson Is Ignoring," *New Statesman*, December 11, 2019, https://www.newstatesman.com /politics/2019/12/it-s-hell-meet-starving-grandfather-outside-tory-hq -who-boris-johnson.

22. I discuss causally acquired special duties to more distant descendants in my article "Do Parents Have a Special Duty to Mitigate Climate Change?"

23. For a moving personal account of the loss of a child, see Angela Miller, August 24, 2016, https://abedformyheart.com/7-things-since-loss-of-child/. A 2019 survey of research confirmed "a significant burden of complicated or prolonged grief in parents of children dying from virtually any cause." See Sue Morris, Kalen Fletcher, and Richard Goldstein, "The Grief of Parents after the Death of a Young Child," *Journal of Clinical Psychology in Medical Settings* 26, no. 3 (2019): 321–338. Another study found long-term mental health problems in mothers and siblings of children with cancer; Jacqui van

Warmerdam et al., "Long-Term Mental Health Outcomes in Mothers and Siblings of Children with Cancer: A Population-Based, Matched Cohort Study," *Journal of Clinical Oncology* 38, no. 1 (2020): 51–62. My argument here also draws on my article "Do Parents Have a Special Duty to Mitigate Climate Change?"

24. Read's book is *Parents for a Future: How Loving Our Children Can Prevent Climate Collapse* (Norwich, UK: UEA Publishing Project, 2021). A former spokesman for Extinction Rebellion, he also calls for mass climate activism by parents. Philosopher Anca Gheaus offers a parallel argument for intergenerational duties. We all, she says, have a right to raise a child we can adequately parent. To secure *this* for the next generation means securing it for the one after, and so on, indefinitely. I'll come back to this later; Gheaus, "The Right to Parent and Duties Concerning Future Generations," *Journal of Political Philosophy* 24, no. 4 (2016): 487–508.

25. The Lambertini quote is from R. E. A. Almond, M. Grooten, and T. Petersen, *Living Planet Report 2020: Bending the Curve of Biodiversity Loss* (Gland, Switzerland: WWF, 2020), 4. The information on extinction rates is from Gerardo Ceballos, Paul R. Ehrlich, and Peter H. Raven, "Vertebrates on the Brink as Indicators of Biological Annihilation and the Sixth Mass Extinction," *Proceedings of the National Academy of Sciences* 117, no. 24 (2020): 13596–13602.

26. I discuss duties of justice to nonhuman individuals, species, and systems in chapter 3 of my book *What Climate Justice Means*. Nussbaum's argument is in chapter 6 of *Frontiers of Justice: Disability, Nationality, Species Membership* (Cambridge, MA, and London: The Belknap Press of Harvard University Press, 2006). On the sophisticated brains of octopi, see Peter Godfrey-Smith, *Other Minds: The Octopus, the Sea, and the Deep Origins of Consciousness* (New York: Farrar, Straus and Giroux, 2016). On mammals' capacities for complex lives, see David J. Mellor, "Updating Animal Welfare Thinking: Moving Beyond the 'Five Freedoms' Towards 'A Life Worth Living,'" *Animals: An Open Access Journal from MDPI* 6, no. 3 (2016), https://doi.org/10.3390/ani6030021.

27. Schlosberg's argument is in part 3 of *Defining Environmental Justice: Theories, Movements and Nature* (New York: Oxford University Press, 2007). The grim prediction for polar bears is from Péter K. Molnár, Cecilia M. Bitz, Marika M. Holland, Jennifer E. Kay, Stephanie R. Penk, and Steven

C. Amstrup, "Fasting Season Length Sets Temporal Limits for Global Polar Bear Persistence," *Nature Climate Change* 10, no. 8 (2020), https://doi.org/10.1038/s41558-020-0818-9.

28. The bulk of conservation dollars goes to species (usually large animals) perceived as attractive, entertaining, or useful. See Ernest Small, "The New Noah's Ark: Beautiful and Useful Species Only. Part 1. Biodiversity Conservation Issues and Priorities," *Biodiversity* 12, no. 4 (2011): 232–247.

29. The sources are Gert-Jan Vanaken and Marina Danckaerts, "Impact of Green Space Exposure on Children's and Adolescents' Mental Health: A Systematic Review," *International Journal of Environmental Research and Public Health* 15, no. 12 (2018), https://doi.org/10.3390/ijerph15122668; Marja I. Roslund et al., "Biodiversity Intervention Enhances Immune Regulation and Health-Associated Commensal Microbiota among Daycare Children," *Science Advances* 6, no. 42 (2020), https://doi.org/10.1126/sciadv.aba2578; Richard Sheldrake, Ruth Amos, and Michael J. Reiss, *Children and Nature: A Research Evaluation for The Wildlife Trusts* (Newark, UK: The Wildlife Trusts; London: UCL Institute of Education; Edinburgh: People's Postcode Lottery, 2019); Helliwell et al., *World Happiness Report 2020*.

30. The Klein quote is from her book *This Changes Everything: Capitalism vs. the Climate* (New York: Penguin), 27. The Hughes tweet is as follows: "I showed the results of aerial surveys of #bleaching on the #GreatBarrierReef to my students. And then we wept." Twitter, April 19, 2016, 8:48 p.m., https://twitter.com/ProfTerryHughes/status/722512223067721728. The findings on sparrows are from Helen Whale and Franklin Ginn, "In the Absence of Sparrows," in *Mourning Nature: Hope at the Heart of Ecological Loss and Grief*, ed. A. Consulo Willox and K. Landman (Montreal: McGill-Queens University Press, 2017), 92–116.

31. Read makes a parallel point. He thinks our care for future generations must extend to nonhuman species (even on purely precautionary grounds). A "*maximally* biodiverse Earth," he says, "is maximally conducive to human health and resilience" (*Parents for a Future*, 74). On Indigenous communities' recognition of this interdependence, see, for example, Kyle Powys Whyte, "Indigenous Women, Climate Change Impacts, and Collective Action," *Hypatia* 29, no. 3 (2014): 599–616.

32. These claims draw on UNEP, Convention on Biological Diversity, and World Health Organisation, *Connecting Global Priorities: Biodiversity and Human Health: A State of Knowledge Review* (World Health Organization and Secretariat of the Convention on Biological Diversity, 2015). The ecofeminist argument is made compellingly in Val Plumwood, *Feminism and the Mastery of Nature* (London and New York: Routledge, 1993). Activist Leah Thomas and (separately) a group of philosophers have expanded the notion of intersectionality to apply across species. See Leah Thomas, "Intersectional Environmentalism Is Our Urgent Way Forward," *Youth to the People*, June 24, 2020, https://www.youthtothepeople.com/blogs /to-the-people/intersectional-environmentalism-is-our-urgent-way-forward; Petra Tschakert et al., "Multispecies Justice: Climate-Just Futures with, for and Beyond Humans," *WIREs Climate Change* 12, no. 2 (2021), https:// doi.org/10.1002/wcc.699.

33. The statistics on antibiotics are from Sriram et al., *State of the World's Antibiotics 2021*, 9. The 80 percent figure is from Almond, Grooten, and Petersen, *Living Planet Report 2020*, 61. The quote is from the World Resources Institute's Global Forest Review: "Deforestation Linked to Agriculture," World Resources Institute, https://research.wri.org/gfr/forest-extent-indicators /deforestation-agriculture. The link between deforestation and pandemic risk is spelled out by Rory Gibb et al., "Zoontic Host Diversity Increases in Human-Dominated Ecosystems," *Nature* 584, no. 7821 (2020): 398–402. See also Yewande Alimi et al., *Report of the Scientific Task Force on Preventing Pandemics* (Cambridge, MA: Harvard Global Health Institute, 2021).

34. For one account of ecological citizenship (and how it diverges from "environmental citizenship"), see Dobson, *Citizenship and the Environment*. For a further, nuanced account, see Sherilyn Macgregor's feminist ecological citizenship: "Only Resist: Feminist Ecological Citizenship and the Post-Politics of Climate Change," *Hypatia* 29, no. 3 (2014); and *Beyond Mothering Earth: Ecological Citizenship and the Politics of Care* (Vancouver: UBC Press, 2014).

35. The Reddit thread is "A Punch in the Gut: Finding Out Your Ancestors Owned Slaves," https://www.reddit.com/r/Genealogy/comments/84flzo/a _punch_in_the_gut_finding_out_your_ancestors/ (accessed November 10, 2022). Davidson tells his story in *The Perfect Nazi: Uncovering My Grandfather's Secret Past* (London: Penguin Books, 2011). However, not everyone

is so concerned. One respondent on the Reddit thread wrote: "Did you own slaves? Did your parents? Grandparents? Anyone who is still alive? I'm guessing not. That being the case, then don't worry about it. You're not responsible for the actions of your ancestors."

CHAPTER 4

1. The comments are from "Introduction/Why Did You Have Children?," Mumsnet, https://www.mumsnet.com/Talk/parenting/1428199-intro duction-why-did-you-have-children (accessed November 10, 2022); "Why Did You Choose to Have Children?," Mumsnet, https://www.mumsnet .com/Talk/parenting/2866687-Why-did-you-choose-to-have-children (accessed November 10, 2022).

2. The unplanned statistics are from Lawrence B. Finer and Mia R. Zolna, "Declines in Unintended Pregnancy in the United States, 2008–2011," *New England Journal of Medicine* 374, no. 9 (2016): 843–852. See D. Langdridge, P. Sheeran, and K. Connolly, "Understanding the Reasons for Parenthood," *Journal of Reproductive and Infant Psychology* 23, no. 2 (2005): 121–133; Abbie Goldberg, Jordan Downing, and April Moyer, "Why Parenthood, and Why Now? Gay Men's Motivations for Pursuing Parenthood," *Family Relations* 61 (2012): 157–174. Eight men in the Goldberg study also specified wanting to shape a child's moral development, especially to encourage tolerance.

3. The Pepino quote is from Stephanie Bailey, "Birthstrike: The People Refusing to Have Kids, Because of 'the Ecological Crisis,'" *CNN*, June 26, 2019. Bickner is quoted in Ted Scheinman, "The Couples Rethinking Kids Because of Climate Change," BBC, September 20, 2019.

4. The "one in four" figure is from Marks et al., "Young People's Voices on Climate Anxiety." For more on the "COVID baby bust," see, for example, Melissa S. Kearney and Philip Levine, "Half a Million Fewer Children? The Coming COVID Baby Bust," Brookings, June 15, 2020, https://www .brookings.edu/research/half-a-million-fewer-children-the-coming-covid -baby-bust/; Francesca Luppi, Bruno Arpino, and Alessandro Rosina, "The Impact of Covid-19 on Fertility Plans in Italy, Germany, France, Spain, and the United Kingdom," *Demographic Research* 43, no. 47 (2020): 1399– 1412; Melissa S. Kearney and Phillip Levine, "Early Evidence of Missing

Births from the COVID-19 Baby Bust," Brookings, December 13, 2021, https://www.brookings.edu/research/early-evidence-of-missing-births -from-the-covid-19-baby-bust/. Since the pandemic, early evidence suggests that births are rebounding. See, for example, Sean Salai, "U.S. Census Sees Birth Rates Rebounding from COVID-19 Pandemic," *Washington Times*, September 24, 2021. "Elena" was quoted in Jennifer Gerson, "How Women Changed Their Minds about Having Kids during the Pandemic," *Bustle*, June 30, 2020.

5. These figures, which are actually comparatively conservative, come from John Halstead and Johannes Ackva, *Climate & Lifestyle Report* (London: Founders Pledge, 2020). According to an earlier study, a US woman increases greenhouse gas emissions by 5.7 times her own lifetime total by having a child; Paul A. Murtaugh and Michael G. Schlax, "Reproduction and the Carbon Legacies of Individuals," *Global Environmental Change* 19 (2009): 14–20. Lund University researchers concluded that having one child fewer than you would have done is by far the most significant carbon-saving action you can take as an individual; Seth Wynes and Kimberly A. Nicholas, "The Climate Mitigation Gap: Education and Government Recommendations Miss the Most Effective Individual Actions," *Environmental Research Letters* 12 (2017), https://doi.org/10.1088/1748-9326/aa7541. However, as Halstead and Ackva point out, this ignores the fact that social changes can cut the per-capita carbon price tag.

6. I talk more about population elsewhere. The short version? "Climate justice means urgently prioritising reproductive justice for its own sake—and welcoming the demographic side effects." See my book *What Climate Justice Means*, 49.

7. The Repugnant Conclusion is the brainchild of philosopher Derek Parfit, *Reasons and Persons*, 1987 corrected reprint ed. (Oxford: Clarendon Press, 1984), 381–391.

8. This argument is borrowed from the philosopher and economist John Broome, "A Reply to My Critics," *Midwest Studies in Philosophy* 40 (2016): 158–171.

9. I discuss this in more detail in my article "Do Parents Have a Special Duty to Mitigate Climate Change?"

10. James's dystopian thriller is *The Children of Men* (London: Faber & Faber, 1992). It was made into a 2006 film by Alfonso Cuaron, to which Read refers to make the same point (*Parents for a Future*, 49).

11. The Ruddick quote is from "Maternal Thinking," 345. Robeyns's article is "Is Procreation Special?," *Journal of Value Inquiry* (2021), https://doi.org/10.1007/s10790-021-09797-y. For the sociological evidence on parenting and happiness, see, for example, Hans-Peter Kohler, Jere R. Behrman, and Axel Skytthe, "Partner + Children = Happiness? The Effects of Partnerships and Fertility on Well-Being," *Population and Development Review* 31, no. 3 (2005): 407–445; Arnstein Aassve, Alice Goisis, and Maria Sironi, "Happiness and Childbearing across Europe," *Social Indicators Research* 108 (2012): 65–86; Thomas Hansen, "Parenthood and Happiness: A Review of Folk Theories Versus Empirical Evidence," *Social Indicators Research*, no. 1 (2011): 29–64.

12. The quote is from "Sonnet 116," in *William Shakespeare: The Complete Works*, ed. Stanley Wells and Gary Taylor (Oxford: Clarendon Press, [1609] 1988). Brighouse and Swift make their argument in *Family Values*, 87–93. Tim Fowler characterizes parenting as a "valuable project" in *Liberalism, Childhood and Justice: Ethical Issues in Upbringing* (Bristol, UK: Bristol University Press, 2021), 107–108. On love and parenting, see also Luara Ferracioli, "Procreative-Parenting, Love's Reasons and the Demands of Morality," *Philosophical Quarterly* 68, no. 270 (2017): 77–97. Anca Gheaus thinks we each have an interest-based *right* to be a parent, at least under normal circumstances; see Gheaus, "Right to Parent."

13. Steinem was interviewed on Elizabeth Day, *S8, Bonus Episode! How to Fail: Gloria Steinem*, podcast audio, 2020. Akbar's article is "Living in a Woman's Body: It's a Potent Myth That All Women Want Children—But I Have Experienced Other Wonders," *The Guardian*, February 10, 2022.

14. The quotes are from "Why Did You Have Kids?," Quora, https://www.quora.com/Why-did-you-have-kids (accessed November 10, 2022); italics in original.

15. The quote is from Stefanie Marsh, "'The Desire to Have a Child Never Goes Away': How the Involuntarily Childless Are Forming a New Movement," *The Guardian*, October 2, 2017. The study surveyed 580 midwestern women; see Julia McQuillan et al., "Frustrated Fertility: Infertility and

Psychological Distress among Women," *Journal of Marriage and Family* 65, no. 4 (2003): 1007–1018. See also, for example, Reija Klemetti et al., "Infertility, Mental Disorders and Well-Being—a Nationwide Survey," *Acta Obstetricia et Gynecologica* 89 (2010): 677–682.

16. Hughes's comment is from Nick Baker and Emma Nobel, "Involuntary Childlessness Can Cause 'Incredible Pain and Grief' for Anyone Who Wants to Be a Parent," ABC News, September 27, 2021.

17. Gheaus's paper is "More Co-parents, Fewer Children: Multiparenting and Sustainable Population," *Essays in Philosophy* 20, no. 1 (2019): 1–21.

18. These philosophers include Daniel Friedrich, "A Duty to Adopt?," *Journal of Applied Philosophy* 30, no. 1 (2013): 25–39; Tina Rulli, "Preferring a Genetically-Related Child," *Journal of Moral Philosophy* 13 (2016): 669–698.

19. The study (mentioned earlier) is Langridge, Sheeran, and Connolly, "Understanding the Reasons for Parenthood." Tina Rulli thinks these *aren't* the kind of reasons that merit moral protection. She compares them to having a child in order to control them, or even to get a financial or emotional hold over the other parent. See Rulli, "Preferring a Genetically-Related Child." Robeyns's argument is in "Is Procreation Special?" For more on having kids as achieving immorality, see Arthur J. Dyck, "Procreative Rights and Population Policy," *The Hasting Center Studies* 1, no. 1 (1973): 74–82. There is also some evidence that there are far fewer "young children in good health" available for adoption than would-be parents looking to adopt them: Nigel Cantwell, "Intercountry Adoption: A Comment on the Number of 'Adoptable' Children and the Number of Persons Seeking to Adopt Internationally," *International Child Protection: The Judges' Newsletter* 5 (2003); Friedrich, "A Duty to Adopt?," 34.

20. For an open and moving account of having twins via surrogacy, see Pragya Agarwal, *Motherhood: On the Choices of Being a Woman* (Edinburgh: Canongate, 2021).

21. These sources are Rosemary Mann et al., "The Personal Experience of Pregnancy for African-American Women," *Journal of Transcultural Nursing* 10, no. 4 (1999): 299; Amy Mullin, *Reconceiving Pregnancy and Childcare: Ethics, Experience, and Reproductive Labor*, Cambridge Studies in Philosophy and Public Policy (Cambridge: Cambridge University Press, 2005).

22. Ocasio-Cortez reportedly made the comment to Instagram followers: see Miranda Green, "Ocasio-Cortez: It's 'Legitimate' to Ask If Ok to Have Children in Face of Climate Change," *The Hill*, February 25, 2019.

23. The quote is from Laura Paddison, "9 People on the Ethics of Having Kids in an Era of Climate Crisis," *HuffPost*, August 8, 2019.

24. Ashcroft's letter is published in the section "This Is How Scientists Feel" of Is This How You Feel? (2020), https://www.isthishowyoufeel.com/this-is-how -scientists-feel.html.

25. Heglar, "Climate Change Isn't the First Existential Threat." Quoted in the introduction.

26. I make this point in my article "Do Parents Have a Special Duty to Mitigate Climate Change?"

27. For Gheaus's argument, see Gheaus, "Right to Parent."

28. McKibben's quote is from *Maybe One: A Case for Smaller Families* (New York: Plume, 1998), 199.

29. The philosopher Chiara Cordelli argues that if we choose freely to take on certain life projects, those projects then can't be used as demandingness reasons not to fulfill duties of beneficence; Cordelli, "Prospective Duties and the Demands of Beneficence," *Ethics* 128, no. 2 (2018): 373–401.

30. The quote is from Sarah Conly, "The Right to Procreation: Merits and Limits," *American Philosophical Quarterly* 42, no. 2 (2005): 107. See also McKibben, *Maybe One*; Rulli, "Preferring a Genetically-Related Child."; Sarah Conly, *One Child: Do We Have a Right to More?* (New York: Oxford University Press, 2015). Conly goes into morally dangerous terrain, saying states should legally *limit* couples to one child. On "stopping at two" see Travis Rieder, *Toward a Small Family Ethic: How Overpopulation and Climate Change Are Affecting the Morality of Procreation* (Cham, Switzerland: Springer, 2016); Christine Overall, *Why Have Children: The Ethical Debate* (Cambridge, MA: MIT Press, 2012); Trevor Hedberg, "The Duty to Reduce Greenhouse Gas Emissions and the Limits of Permissible Procreation," *Essays in Philosophy* 20, no. 1 (2019). I assume that there is significant value in having a sibling, but it's harder to claim that this is a fundamental human interest. In 1998, McKibben surveyed the evidence that only children were

disadvantaged or socially damaged and found it to be over-reliant on one piece of dubious empirical research (*Maybe One*, 44–62).

31. Agarwal discusses this in *Motherhood*.

32. For Sasser's full argument, see *On Infertile Ground: Population Control and Women's Rights in the Era of Climate Change* (New York: NYU Press, 2018).

CHAPTER 5

1. For more on these mother-activists, see Alexis Jetter, "Patsy Ruth Oliver: A Mother's Battle for Environmental Justice," *Journal of the Motherhood Initiative for Research and Community Involvement* 3, no. 2 (2001): 89–96; Melanie Panitch, *Disability, Mothers, and Organisation: Accidental Activists* (New York and London: Routledge, 2008); "Mothers Demand Climate Action in London March," BBC News, May 12, 2019.

2. The International Energy Agency quote is from IEA, *Renewables 2021: Analysis and Forecast to 2026* (Paris: IEA, 2021), 20. The report's forecasts for renewable capacity growth were 80 percent lower than needed for net zero by 2050; annual demand growth for biofuels needs to quadruple. For more on the scale of technological transformation needed to tackle climate change, and the timescale of previous ones, see UN DESA, *World Economic and Social Survey 2011: The Great Green Technological Transformation* (New York: United Nations, 2011).

3. The three strands on tackling antibiotic resistance are from Steven J. Hoffman and Kevin Outterson, "What Will It Take to Address the Global Threat of Antibiotic Resistance?," *Journal of Law, Medicine & Ethics* 43, no. 2: (2015): 363–368.

4. On the UK's asylum-seeker policy, see "One-Way Ticket to Rwanda for Some UK Asylum Seekers," BBC News, April 14, 2022. Also as I write, the United States (and the world) is reeling from another mass murder in an elementary school. See Tim Swift, "'Thoughts and Prayers Are Not Enough:' Md. Leaders Urge Gun Control after Texas Shooting," *Fox New*, April 14, 2022. Farhana Sultana makes the case for climate reparations in "Critical Climate Justice," *Geographical Journal* 188, no. 1 (2021): 118–124. After the 2021 26th Conference of Parties in Glasgow, Climate Action Tracker warned that the 2030 nationally determined contribution (NDC) targets

would mean temperature increases of 2.4°C on preindustrial levels by 2100. Even if states stuck to their NDCs and all announced longer-term targets, it would be 1.8°C. See "Glasgow's 2030 Credibility Gap: Net Zero's Lip Service to Climate Action" (Berlin: Climate Analytics and New Climate Institute, 2021). I discuss the Paris Agreement in my book *What Climate Justice Means*, 129–138.

5. The sources are Climate Accountability Institute, "Update of Carbon Majors 1965–2018," December 9, 2020. For a discussion of the role played by big food producers in worsening greenhouse gas emissions, see Oliver Lazarus, Sonali McDermid, and Jennifer Jacquet, "The Climate Responsibilities of Industrial Meat and Dairy Producers," *Climatic Change* 165, no. 1 (2021). The information on deforestation is from ZSL, "Companies Failing to Protect Millions of Hectares of Tropical Forest," news release, July 17, 2020. For information on child labor, see for example SOMO, "Fact Sheet: Child Labour in the Textile and Garment Industry" (Amsterdam: SOMO, 2014). The report on the climate impact of corporate cash is James Vaccaro and Paul Moinester, *The Carbon Bankroll: The Climate Impact and Untapped Power of Corporate Cash* (Washington: Climate Safe Lending Network, The Outdoor Policy Outfit, Bank FWD, 2022). Including finance emissions increased PayPal's carbon emissions by 5,512 percent; Disney's by 169 percent.

6. The quote is from Meredith Niles, associate professor in food systems at the University of Vermont, Zoom interview with the author, August 24, 2021. Fossil fuel subsidies fell to $180bn in the pandemic, but were back to a horrifying $440bn in 2021. See the International Energy Agency, "Energy Subsidies: Tracking the Impact of Fossil Fuel Subsidies," https://www.iea .org/topics/energy-subsidies (accessed August 19, 2022).

7. The UK cabinet advisor is Nicky Brennan; the quote is from Francesca Cambridge Mallen, *School Uniform: Dressing Girls to Fail* (London: Let Clothes Be Clothes, 2021), 4. For more on the need to address racism in law enforcement, see Black Lives Matter, "BLM Demands," https://black livesmatter.com/blm-demands/ (accessed November 10, 2022).

8. For Gardiner's argument, see "A Call for a Global Constitutional Convention Focused on Future Generations," *Ethics & International Affairs* 28, no. 3 (2014): 299–315.

9. All quotes from Brummell are from a Zoom interview with the author on September 2, 2021.

10. The statistics on plastics are from Hannah Ritchie and Max Roser, "Plastic Pollution" (Oxford: Oxford Martin School & Global Change Data Lab, 2018), https://ourworldindata.org/plastic-pollution.

11. The sources for these claims are Beth Turnbull, Melissa L. Graham, and Ann R. Taket, "Social Exclusion of Australian Childless Women in Their Reproductive Years," *Social Inclusion* 4, no. 1 (2016), https://doi.org/10.17645/si.v4i1.489; Tim Huijts, Gerbert Kraaykamp, and S. V. Subramanian, "Childlessness and Psychological Well-Being in Context: A Multilevel Study on 24 European Countries," *European Sociological Review* 29, no. 1 (2013): 32–47; Jessica Autumn-Brown and Myra Marx Ferree, "Close Your Eyes and Think of England: Pronatalism in the British Print Media," *Gender & Society* 19, no. 1 (2005): 5–24. In 2022, the pope added fuel to the fire: Harriet Sherwood, "Choosing Pets over Babies Is 'Selfish and Diminishes Us,' Says Pope," *The Guardian*, January 5, 2022.

12. The Berners-Lee quote is from *There Is No Planet B: A Handbook for the Make or Break Years* (Cambridge: Cambridge University Press, 2019), 171.

13. See Climate Families NYC (@ClimateFamsNYC), "BREAKING. This weekend, we paid a second visit to @blackrock CEO Larry Fink's country mansion . . . ," Twitter, May 16, 2022, 1:03 p.m., https://twitter.com/Climate FamsNYC/status/1526246990887235584.

14. Unless otherwise specified, all Mailer quotes are from a Zoom interview with the author on September 9, 2021.

15. See Mothers Rise Up (@mothersriseup), "UPDATE. Yesterday, with @parents_4future, we met with Bruce Carnegie-Brown @BNCB Chair of @LloydsofLondon and dad of four . . . ," Twitter, May 18, 2022, 10:10 a.m. https://twitter.com/mothersriseup/status/1526928323263307777. For a video of the *Mary Poppins* protest, see Mothers Rise Up (@mothersriseup), "BREAKING: Dancers, parents & kids form flash mob outside @Lloydsof London to urge the insurer to stop enabling dangerous fossil fuels . . . ," Twitter, June 13, 2022, 7:00 a.m., https://twitter.com/mothersriseup/status/1536302489707941889. The film in question is *Mary Poppins*, directed by Robert Stevenson (Walt Disney, 1964).

16. The sources on market size are IMARC, "Diaper Market: Global Industry Trends, Share, Size, Growth, Opportunity and Forecast 2022–2027," IMARC Group, https://www.imarcgroup.com/prefeasibility-report-diaper-manufacturing-plant-2 (accessed August 19, 2022); Raju Kale and Roshan Deshmukh, *Baby Food Market by Product Type (Dried Baby Food, Milk Formula, Prepared Baby Food, and Others), and Distribution Channel (Supermarket, Hypermarket, Small Grocery Retailers, Health and Beauty Retailers, and Others): Global Opportunity Analysis and Industry Forecast, 2021–2027* (Portland, OR: Allied Market Research, 2020); Research and Markets, *Children's Wear: Global Market Trajectory & Analytics* (New York: Research and Markets, 2021).

17. Chenoweth and Stephan's book is *Why Civil Resistance Works: The Strategic Logic of Nonviolent Conflict* (New York: Columbia University Press, 2012). The information on parents in the United States is from United States Census Bureau, "Census Bureau Releases New Estimates on America's Families and Living Arrangements," news release, December 2, 2020, https://www.census.gov/newsroom/press-releases/2020/estimates-families-living-arrangements.html.

18. The quote is from Maya Mailer, "Why Mothers Are Taking the Fight for Climate Action to Lloyd's of London," Oxfam blog, May 26, 2022, https://frompoverty.oxfam.org.uk/why-mothers-are-taking-the-fight-for-climate-action-to-lloyds-of-london/.

19. For Trachtenberg's argument, see "Complex Green Citizenship and the Necessity of Judgement," *Environmental Politics* 19, no. 3 (2010): 339–355.

20. On expected consequences, see Derek Parfit, *Reasons and Persons*. The arguments of this section (as well as the discussion in chapter 7 of individual emissions cuts and difference-making) are made more briefly in my book *What Climate Justice Means*, 153–183. They are based on my academic paper "The Cooperative Promotional Model & Its Challenges," in *Climate Justice and Non-state Actors: Corporations, Regions, Cities, and Individuals*, ed. Jeremy Moss and Laclan Umbers (London: Routledge, 2020), 101–117.

21. Nefsky's paper is "How You Can Help, without Making a Difference," *Philosophical Studies*, no. 174 (2017): 2743–2767.

22. For Cullity's argument, see "Moral Free Riding," *Philosophy and Public Affairs* 24, no. 1 (1995): 3–34.

23. The figures are from "21x: It's the Most Powerful Thing You Can Do to Protect the Planet," *Make My Money Matter*, https://makemymoneymatter.co.uk/21x/ (accessed November 10, 2022); Rainforest Action Network et al., *Banking on Climate Chaos: Fossil Fuel Finance Report 2021* (2021), https://www.ran.org/wp-content/uploads/2021/03/Banking-on-Climate-Chaos-2021.pdf.

24. The Children's Parliament initiative uses rights-based, creative practice to help marginalized children's needs and experiences influence national legislation. See "About Us," Children's Parliament, https://www.childrensparliament.org.uk/about-us/ (accessed November 10, 2022).

25. I talk about this more in Cripps, "Is Civil Disobedience OK If It's the Only Way to Prevent Climate Catastrophe?," *The Guardian*, April 12, 2022.

26. This is inspired by philosopher Donald Regan's "cooperative utilitarianism," but it's the idea, not the name, that matters. For a detailed academic version of the argument, see Cripps, "The Cooperative Promotional Model & Its Challenges," and Elizabeth Cripps, "Intergenerational Ethics and Individual Duties: A Cooperative Promotional Approach," in *The Oxford Handbook of Intergenerational Ethics*, ed. Stephen Gardiner (Oxford: Oxford University Press, 2021), https://doi.org/10.1093/oxfordhb/9780190881931.013.42. Of course, this isn't the only option. I could think like a rule consequentialist: start with what *everyone* should do to bring about the best overall result, then do my "bit" of that. But this suffers from the serious flaw that many other people, including other parents, won't do anything without political, corporate, or norm change to nudge them along the way.

27. The campaign group Survival International has documented instances of conservation colonialism: "Decolonize Conservation: Indigenous People Are the Best Conservationists," https://www.survivalinternational.org/conservation (accessed November 10, 2022). I discuss this in my book *What Climate Justice Means*, 90–91.

28. For Feeney's story, see Steven Bertoni, "Exclusive: The Billionaire Who Wanted to Die Broke . . . Is Now Officially Broke," *Forbes*, September 15, 2020. That's not to say that the causes to which Feeney donated his vast fortune are the ones best suited to tackle global emergencies.

29. This was on social media: Prof. Katharine Hayhoe (@KHayhoe), "Every year, I add 2 new low-carbon habits to my life. But every DAY, I do the

most impt thing anyone can . . . ," Twitter, January 26, 2021, 12:50 p.m., https://twitter.com/KHayhoe/status/1354124680978968576. See also her excellent book *Saving Us: A Climate Scientist's Case for Hope and Healing in a Divided World* (Atria: One Signal Publishers, 2021).

30. "Ask them if there's a real need for you or your child to have an antibiotic," says Otto Cars. "Unlock a dialogue with them" (Zoom interview with the author, September 22, 2021).

31. For Hayhoe's argument, see "How to Talk About Climate Change So People Will Listen," *Chatelaine*, April 18, 2019; see also Hayhoe, *Saving Us.*

32. The study is Seth Wynes, Matthew Motta, and Simon Donner, "Understanding the Climate Responsibility Associated with Elections," *One Earth* 4 (2021): 363–371.

33. I was looking at "A Couple at the SF Pride Parade Today," imgur, June 28, 2015, https://imgur.com/zdv2Mk5.

34. The views of young Republicans are from Cary Funk and Meg Hefferon, "U.S. Public Views on Climate and Energy," Pew Research Center, November 25, 2019, https://www.pewresearch.org/science/2019/11/25/u-s-public -views-on-climate-and-energy/.

35. The quote from Backer is from a Zoom interview with the author on October 11, 2021.

36. The story and the quote below are from Harriet Grant, "'Don't Make This Rich V Poor': The Mothers Who Won the Right for Their Kids to Play Together," *The Guardian*, March 30, 2019.

37. Lorde was giving a keynote talk at the National Women's Studies Association Conference in 1981. It is reproduced as Lorde, "The Uses of Anger," *Womens Studies Quarterly* 9 no. 3 (1981): 7–10. The O'Keefe quote is from Emily Stewart, "How to Be a Good White Ally, According to Activists," *Vox*, June 2, 2020.

38. See Laura Varley (@rocketminx), "@EverydaySexism guy at a hotel gave me 'that look' and said 'helloooo' to me. His friend said: 'Don't be a dick' – need more guys like that," Twitter, March 12, 2014, 3:38 p.m., https:// twitter.com/rocketminx/status/443833384536928256. For more examples, see Laura Bates, "The Men Who Help Fight Back against Everyday Sexism," *The Guardian*, March 14, 2014. The sociologists' advice is in Meredith Nash

et al., "'It's Not about You': How to Be a Male Ally," *The Conversation*, April 25, 2021, https://theconversation.com/its-not-about-you-how-to-be -a-male-ally-158134.

CHAPTER 6

1. I can't claim originality for these words: they are an amalgam of suggestions from various threads or discussions on "positive parenting," absorbed over the years.

2. Al Gore's documentary film is directed by Davis Guggenheim, *An Inconvenient Truth* (2006). For an anti-racist perspective, see, for example, Blake Thorkelson, "Stay Hopeful and Do 'Uncomfortable Things' Advises Justice Advocate Bryan Stevenson," *Yale News*, February 2, 2017.

3. All references to Mulnix and Mulnix concern *Happy Lives, Good Lives*.

4. See Aristotle, "Nicomachean Ethics"; Plato, *The Republic*, trans. B. Jowett (Project Gutenberg, [375 BC] 1998). Nussbaum's "capabilities approach" is a modern Aristotelian account; see Nussbaum, *Women and Human Development*. Not all philosophers agree that happiness or well-being requires acting morally. An exception is Nozick, *The Examined Life*. For a detailed discussion, see Mulnix and Mulnix, *Happy Lives, Good Lives*; or Hurka, *The Best Things in Life*.

5. The quote is from Mulnix and Mulnix, *Happy Lives, Good Lives*, 30.

6. The Pew Research Center study is "Teaching the Children: Sharp Ideological Differences, Some Common Ground," Pew Research Center, September 18, 2014, https://www.pewresearch.org/politics/2014/09/18/teaching-the -children-sharp-ideological-differences-some-common-ground/.

7. This argument is from my article "Justice, Integrity, and Moral Community: Do Parents Owe It to Their Children to Bring Them up as Good Global Climate Citizens?," *Proceedings of the Aristotelian Society* 117, no. 1 (2017): 41–59.

8. The Harvey quote is from *Raising White Kids*, 40.

9. I'm using Joseph Raz's influential account of autonomy; Raz, *The Morality of Freedom* (Oxford: Clarendon Press, 1986). As discussed, Nussbaum also emphasizes forming one's own life goals. See Nussbaum, *Women and Human Development*, 79. Legal theorist Joel Feinberg stresses the importance of

"an open future": "The Child's Right to an Open Future," in *Whose Child? Children's Rights, Parental Authority, and State Power*, ed. William Aiken and Hugh LaFollette (Totowa, NJ: Littlefield, Adams, & Co., 1980), 124–153.

10. For Harry Brighouse and Adam Swift, doing things you value together is a core "relationship good." See Brighouse and Swift, *Family Values*, 124.

11. Brennan's article is "The Goods of Childhood and Children's Rights," in *Family-Making: Contemporary Ethical Challenges*, ed. Francoise Baylis and Caroline McLeod (Oxford: Oxford University Press, 2014), 29–44.

12. Betjeman's autobiography is *Summoned by Bells* (London: John Murray, [1960] 2007), 33.

13. The poems are Helen Dunmore, "To My Nine-Year-Old Self," in *Glad of These Times* (Hexham, UK: Bloodaxe Books, 2014), 55; Emily Dickinson, "The Child's Faith Is New," in *Hope Is the Thing with Feathers: The Complete Poems of Emily Dickinson* (Gibbs Smith). The quote that follows is from Dickinson's poem. The Brennan quote is from "The Goods of Childhood and Children's Rights," 43. References to Colin MacLeod in this section are from his "Agency, Authority and the Vulnerability of Children," in *The Nature of Children's Well-Being: Theory and Practice*, ed. Alexander Bagattini and Colin M. MacLeod (Dordrecht: Springer, 2015), 59, and his "Primary Goods, Capabilities, and Children," in *Measuring Justice: Primary Goods and Capabilities*, ed. Harry Brighouse and Ingrid Robeyns (Cambridge: Cambridge University Press, 2010), 174–192.

14. The Fowler quote is from *Liberalism, Childhood and Justice*, 130. Fowler is avowedly "perfectionist" in his approach to education. Brighouse and Swift are not, but still think parents have a "fiduciary responsibility" for their children's moral, as well as their cognitive, development (*Family Values*, 70).

15. Thanks to Tim Fowler for making me think more about this point. Brighouse and Swift talk more about finding that difficult balance in *Family Values*.

16. For Clayton's argument, see *Justice and Legitimacy in Upbringing*.

17. MacLeod makes the point about "fast-tracking" in "Agency, Authority and the Vulnerability of Children." Following COP26, climate scientist Chip Fletcher wrote a comment piece whose title said it all: "Cop26 Has Failed

Our Children—Political Compromise Cannot Be the Answer," *The Hill*, November 17, 2021.

18. The term "anticipatory parenting" is from Alice Baderin's unpublished paper "'The Talk': Risk, Racism and Family Relationships." Hannan focuses on sexual innocence; see "Why Childhood Is Bad for Children," *Journal of Applied Philosophy* 35, no. S1 (2018): 15–16. For her, childhood is so characterized by vulnerability, imperfect ability to reason, and the need to be dominated that it's a bad state to be in. I disagree, but agree on the ambiguity of some childhood goods.

19. Quotes from Murungi are all from a Zoom interview with the author, September 26, 2021. The debate on a child's "right to a voice" is philosophically complex. David Archard and Suzanne Uniacke unpick this in "The Child's Right to a Voice," *Res Publica* 4 (2020): 1–16. Philosophers Paul Bou-Habib and Serena Olsaretti discuss children's developing autonomy, rejecting the view that it is only their *future* (adult) autonomy that is valuable: "Autonomy and Children's Well-Being," in *The Nature of Children's Well-Being: Theory and Practice*, ed. Alexander Bagattini and Colin Macleod (Dordrecht: Springer Netherlands, 2015), 15–33.

20. See Harvey, *Raising White Kids*, 27. Harvey credits psychologist Beverley Daniel Tatum for the analogy. (For another discussion of how early kids can internalize different ways of responding to race, see the same chapter.) See also Pragya Agarwal, *Wish We Knew What to Say: Talking with Children About Race* (London: Dialogue Books, 2020).

21. The Fowler quote is from *Liberalism, Childhood and Justice*, 129.

22. The Newsome quote is from Tawny "My Name is Tawny" Newsome (@ TrondyNewman), "For anyone worried about talking to their white children right now . . . ," Twitter, June 2, 2020, 7:51 p.m., https://twitter.com/trondynewman/status/1267967053639479296. For Tamir Rice's mother 's story, see: Samaria Rice, "My 12-Year-Old Son, Tamir Rice, Was Killed by Police. I'm Not Allowed to Be Normal," ABC News, July 13, 2020. Shabazz's experience is from "What Black Parents Have to Teach Their Kids That White Parents Don't," Care.com, July 15, 2020, https://www.care.com/c/what-black-parents-must-teach-kids/. Luke's blogpost is "Racism and My Mental Health," November 25, 2020, https://www.youngminds.org.uk/young-person/blog/racism-and-my-mental-health/.

23. The Harvey quote is from *Raising White Kids*, 138.

24. The Harvey quote is from *Raising White Kids*, 62. The Moyer quote is from *How to Raise Kids Who Aren't Assholes: Science-Based Strategies for Better Parenting—from Tots to Teens* (London: Headline Home, 2021), 101.

25. The psychologists' quote is from Association of Clinical Psychologists UK, "ACP-UK Supports the Call for Clinical Psychologists to Speak out on Climate Change," *ACP-UK*, September 17, 2019. The US data is from Liz Hamel et al., "The Kaiser Family Foundation/Washington Post Climate Change Survey" (San Francisco: Kaiser Family Foundation, 2019). The COVID-19 impact was confirmed by a 2021 survey of 116 peer-reviewed and preprint articles: Hasina Samji et al., "Review: Mental Health Impacts of the COVID-19 Pandemic on Children and Youth—a Systematic Review," *Child and Adolescent Mental Health* 27, no. 2 (2022): 173–189. The UK figures are from Richard Atherton, "Climate Anxiety: Survey for BBC Newsround Shows Children Losing Sleep over Climate Change and the Environment," *BBC Newsround*, March 3, 2020. Sharma's op-ed is: "Dear Politicians, Young Climate Activists Are Not Abuse Victims, We Are Children Who Read News," *The Guardian*, April 28, 2022. For more on this, see the work of Caroline Hickman, including: "We Need to (Find a Way to) Talk About . . . Eco-Anxiety," *Journal of Social Work Practice* 34, no. 4 (2020): 411–424.

26. All quotes from Susan Clayton are from a Zoom interview with the author on September 10, 2021.

27. See Harvey, *Raising White Kids*, 115.

28. Agarwal's post was Prof. Pragya Agarwal (@DrPragyaAgarwal), "Just some thoughts on how to talk to young children about war and conflict: be honest, explain and don't brush away their questions . . . ," Twitter, February 26, 2022, 1:28 p.m., https://mobile.twitter.com/DrPragyaAgarwal/status/1497639754577465351.

29. The Shugarman quote is from her excellent book *How to Talk to Your Kids About Climate Change: Turning Angst into Action* (Gabriola Island, Canada: New Society Publishers, 2020), 95.

30. Murungi's books are Maria Horne, Herbert Murungi, and Robert H. Horne, *James the Steward* (Independently Published, 2020); Herbert Murungi,

Maria Horne, and Robert H. Horne, *Keeper of the Forest* (Independently Published, 2021). On nature and cultivating environmental behavior, see Louise Chawla and Victoria Derr, "The Development of Conservation Behaviors in Childhood and Youth," in *The Oxford Handbook of Environmental and Conservation Psychology*, ed. Susan D. Clayton (Oxford and New York: Oxford University Press, 2012), 527–555.

31. The first in these series are Elena Favilli and Francesca Cavallo, *Goodnight Stories for Rebel Girls* (London: Penguin, 2017); Adam Gidwitz, *The Creature of the Pines (The Unicorn Rescue Society)* (New York: Dutton Children's Books, 2018); Asia Citro, *Dragons and Marshmallows (Zoey and Sassafras)* (Woodinville, WA: Innovation Press, 2017).

32. See Charles Dickens, *Hard Times* (London: HarperCollins, [1854] 2012).

33. On authoritative versus authoritarian parenting, see Deborah Laible et al., "The Socialization of Children's Moral Understanding in the Context of Everyday Discourse," in *The Oxford Handbook of Parenting and Moral Development*, ed. Deborah Laible, Gustavo Carlo, and Laura Padilla-Walker (Oxford: Oxford University Press, 2019), 287–298; Laura M. Padilla-Walker and Daye Son, "Productive Parenting and Moral Development," in *The Oxford Handbook of Parenting and Moral Development*, 301–313.

34. The Lewis and Franz study is "Mothers' and Adolescents' Conversations About Volunteering: Insights into the Reasons and Socialization Processes," in *The Oxford Handbook of Parenting and Moral Development*, 269–284. The study is Jason M. Cowell and Jean Decety, "Precursors to Morality in Development as a Complex Interplay between Neural, Socioenvironmental, and Behavioral Facets," *Proceedings of the National Academy of Sciences* 112, no. 41 (2015): 12657–12662.

35. The Ray quote is from *A Field Guide to Climate Anxiety*, 72.

36. The age-appropriate advice here draws (too briefly!) on the detailed and very useful accounts of climate education, anti-racist education, and raising feminist boys, in (respectively) Shugarman, *How to Talk to Your Kids About Climate Change*; Agarwal, *Wish We Knew What to Say*; Bobbi Wegner, *Raising Feminist Boys: How to Talk with Your Child About Gender, Consent and Empathy* (Oakland, CA: Harbinger Publications, 2021).

37. The quote is from Wegner, *Raising Feminist Boys*, 161.

38. The Lapsley quote is from "Moral Formation in the Family: A Research Agenda in Time Future," in *The Oxford Handbook of Parenting and Moral Development*, 394–413. See also Laura M. Padilla-Walker and Katherine J. Christensen, "Empathy and Self-Regulation as Mediators between Parenting and Adolescents' Prosocial Behavior toward Strangers, Friends, and Family," *Journal of Research on Adolescence* 21, no. 3 (2011): 395–396.

39. The sources are Rosemary Randall and Paul Hoggett, "Engaging with Climate Change: Comparing the Cultures of Science and Activism," in *Climate Psychology: On Indifference to Disaster*, ed. Paul Hoggett (Cham, Switzerland: Palgrave Macmillan, 2019), 239–261; Robert Tollemache, "We Have to Talk About . . . Climate Change," in Hoggett, *Climate Psychology*, 217–237.

40. The sources are Agarwal, *Wish We Knew What to Say*, 92–23; A. Cheskis et al., "Americans Support Teaching Children about Global Warming," Yale Program on Climate Change Communication, April 11, 2018, https://climatecommunication.yale.edu/publications/americans-support-teaching-children-global-warming/; Rashawn Ray and Alexandra Gibbons, "Why Are States Banning Critical Race Theory?," Brookings, November 2021, https://www.brookings.edu/blog/fixgov/2021/07/02/why-are-states-banning-critical-race-theory/.

41. For the vice president elect's full speech, see Kamala Harris, "'You Chose Truth': Kamala Harris's Historic Victory Speech in Full—Video," *The Guardian*, November 8, 2020.

CHAPTER 7

1. The sources are Climate Accountability Institute, "Carbon Majors: Update of Top Twenty Companies 1965–2018," Climate Accountability Institute press release, December 9, 2020, https://climateaccountability.org/pdf/CAI%20PressRelease%20Dec20.pdf; World Resources Institute, "Historical GHG Emissions," ClimateWatch, https://www.climatewatchdata.org/ghg-emissions (accessed August 22, 2022); Geoffrey Supran and Naomi Oreskes, "Rhetoric and Frame Analysis of ExxonMobil's Climate Change Communications," *One Earth* 4, no. 5 (2021): 696–719.

2. According to Founders Pledge, donating $1,000 a year could cut greenhouse gases by 100 tonnes, by reducing deforestation or developing clean

technology. That's equivalent to avoiding 62.5 transatlantic flights. See Halstead and Ackva, *Climate & Lifestyle Report*. A recent study found that switching to a "green" pension would, on average, cut carbon twenty-one times more than going vegetarian, quitting flying, *and* changing to a renewable energy provider. See "Pension Fund Carbon Savings Research: A Summary of the Approach," Make My Money Matter campaign, WWF, Aviva (2022), https://www.aviva.com/sustainability/communities/powerofyourpension/. For a description of the so-called Jevons Paradox, see Berners-Lee, *No Planet B*, 82–83.

3. The expected GHG impact of having a child is from Halstead and Ackva, *Climate & Lifestyle Report*. Shailee Koranne criticizes the Birthstriker movement for (among other concerns) implicitly accepting an "individualisation" of responsibility; Koranne, "Why 'Birth Strikes' Aren't the Right Way to End the Climate Crisis," Prism, February 10, 2020, https://prismreports .org/2020/02/10/why-birth-strikes-arent-the-right-way-to-end-the-climate -crisis/.

4. Mainstream environmentalism (e.g., the zero-waste movement) can ignore or even exploit unjust gender norms. See Joseph Murphy and Sarah Parry, "Gender, Households and Sustainability: Disentangling and Re-entangling with the Help of 'Work' and 'Care,'" *Environment and Planning E: Nature and Space* 4, no. 3 (2021): 1099–1120.

5. See Kendi's *How to Be an Antiracist* (New York: Vintage, 2019). White woman Amy Cooper called police on an African American, Christian Cooper, who was birdwatching in Central Park, after he asked her to put her dog on a lead, required in the Ramble portion of the park. See Jan Ransom, "Amy Cooper Faces Charges after Calling Police on Black Bird-Watcher," *New York Times*, October 14, 2020.

6. This quote is from Nash et al., "'It's Not about You.'"

7. The first quote is from Nash et al., "It's Not about You." The O'Keefe quotes are from Stewart, "How to Be a Good White Ally."

8. The Wynes and Nicholas study is "The Climate Mitigation Gap."

9. Take the Jump's advice is from "Less Stuff, More Joy," Take the Jump, https:// takethejump.org/. The Founders Pledge data is from Halstead and Ackva, *Climate & Lifestyle Report*.

10. The Berners-Lee quote is from *No Planet B*, 21. In 2018, the UK's chief medical officer, Dame Sally Davies, urged consumers to buy higher welfare or organic meat to combat antibiotic resistance; Bethan Staton, "Chief Medical Officer Urges Public to Buy Organic Meat," *Sky News*, November 19, 2018.

11. The statistics are from J. Poore and T. Nemecek, "Reducing Food's Environmental Impacts through Producers and Consumers," *Science* 360, no. 6392 (2018): 987–992. How meat is produced also matters. The highest-impact quarter of producers accounts for 56 percent of global emissions and 61 percent of land use for beef from beef herds. The most efficient producers are responsible for *less* greenhouse gas per 100 grams of protein than for some lamb or mutton, cheese, crustaceans, pork, or poultry.

12. The quote is from Berners-Lee, *No Planet B*, 33; italics in original.

13. The statistics are from FAO, "Food Loss and Food Waste," Food and Agriculture Organization of the United Nations, https://www.fao.org/nutrition/capacity-development/food-loss-and-waste/en/ (accessed August 22, 2022).

14. The discussion of virtue ethics draws on Rosalind Hursthouse, "Environmental Virtue Ethics," in *Working Virtue: Virtue Ethics and Contemporary Moral Problems*, ed. Rebecca L. Walker and Philip J. Ivanhoe (Oxford: Oxford University Press, 2007), 155–209. Environmental philosopher Dale Jamieson thinks we should learn to think like virtue theorists, to motivate ourselves to do what would be best overall. See Jamieson, "When Utilitarians Should Be Virtue Theorists," *Utilitas* 19, no. 2: 160–183. Marion Hourdequin makes the point about integrity in "Climate, Collective Action, and Individual Ethical Obligations," *Environmental Values* 19 (2010): 443–464.

15. Kagan's paper is "Do I Make a Difference?," *Philosophy and Public Affairs* 39, no. 2 (2011): 105–141.

16. John Broome makes the point that our individual emissions could trigger extreme weather in "Against Denialism," *The Monist* 102, no. 1 (2019): 110–129.

17. Philosophically, the appeals to "helping to harm" and free riding are based (respectively) on Nefsky, "How You Can Help, without Making a Difference"; Cullity, "Moral Free Riding" and "Climate Harms."

18. I make the point about perceived hypocrisy in my book *Climate Change and the Moral Agent*.

19. The studies on behavior change include Jennifer Long, Niki Harré, and Quentin D. Atkinson, "Understanding Change in Recycling and Littering Behavior across a School Social Network," *American Journal of Community Psychology* 53, no. 3–4 (2014): 462–474; Nicholas A. Christakis and James H. Fowler, "The Collective Dynamics of Smoking in a Large Social Network," *New England Journal of Medicine* 358, no. 21 (2008): 2249–2258. Sociologists studied a social network of 12,067 people, using the 1971–2003 Framingham Heart Study. All quotes from Kateman are from a Zoom interview with the author on September 9, 2021.

20. For more empirical evidence on lifestyle and policy change, see Shahzeen Z. Attari, David H. Krantz, and Elke U. Weber, "Climate Change Communicators' Carbon Footprints Affect Their Audience's Policy Support," *Climatic Change* 154, no. 3 (2019): 2249–2258. Revenue from electric cargo bikes is predicted to grow three times over by 2031, taking it to $1.9 billion worldwide. See Persistence Market Research, "Electric Cargo Bike Market," https://www.persistencemarketresearch.com/market-research/electric-cargo-bikes-market.asp (accessed August 18, 2022). For a detailed discussion of citizen-consumers, see G. Spaargaren and P. Oosterveer, "Citizen-Consumers as Agents of Change in Globalizing Modernity: The Case of Sustainable Consumption," *Sustainability* 2, no. 7 (2010): 1887–1908.

21. For an account of the rise of vegan eating, see Dan Hancox, "The Unstoppable Rise of Veganism: How a Fringe Movement Went Mainstream," *The Guardian*, April 1, 2018. My colleague Mihaela Mihai, political theorist, worries that the Birthstrikers compound problematic heteronormative and pro-natalist norms, in "Ecological Guilt, Political Mourning and Contestatory Citizenship: Responsibility and Its Ambiguities" (unpublished).

22. Butt's paper is "Corrupting the Youth: Should Parents Feed Their Children Meat?," *Ethical Theory and Moral Practice* 24 (2021): 981–997.

23. Williams's argument is in *Moral Luck*.

CHAPTER 8

1. Pre-pandemic, psychologists studied 17,409 parents in forty-two countries, looking for emotional exhaustion, contrast with previous self, loss of pleasure in the parental role, and emotional distancing from children. See Isabelle Roskam et al., "Parental Burnout around the Globe: A 42-Country

Study," *Affective Science* 2, no. 1 (2021): 58–79. In the United States, psychologists surveyed 433 parents (mostly mothers) in May and September 2020: Elizabeth L. Adams et al., "Parents Are Stressed! Patterns of Parent Stress Across COVID-19," *Frontiers in Psychiatry* 12, no. 300 (2021), https://doi.org/10.3389/fpsyt.2021.626456. The Co-Space Oxford study examined six thousand UK parents over 2020: Adrienne Shum et al., *Report 07: Changes in Parents' Mental Health Symptoms and Stressors from April to December 2020* (Oxford: Co-Space, 2021). The disproportionate impact on women has been widely demonstrated, including in Xue and McMunn, "Gender Differences in Unpaid Care Work."

2. My sources are Jacob Geanous, "Brave Widowed Mother Died Saving Her Five Children from House Fire," *Metro*, February 13, 2019; Sara Sanchez, "El Paso Mom Killed in Tippin Elementary Accident Was Army Veteran, Nurse," *El Paso Times*, August 14, 2018; Stabroek News, "Jamaican Mother Dies Trying to Rescue Children from Fire," *Stabroek News*, June 2, 2020; Kim Moodie, "Northland Mother Died Saving Her Child from a Rip," *New Zealand Herald*, February 23, 2020; BBC, "Telford Father Jonathan Stevens 'Died a Hero Saving Children,'" BBC, August 3, 2020.

3. See Epstein, "What Makes a Good Parent?" (The other predictor from the "top three," alongside love and affection, was parents' relationship with each other.)

4. Dickens's character is *Mrs.* Jellyby and his portrayal is deeply gender-stereotyped; see *Bleak House* (Ware, Herts.: Wordsworth, 1853).

5. This is a variant of Matthew Clayton's argument in *Justice and Legitimacy in Upbringing*.

6. See Swift, *How Not to Be a Hypocrite*.

7. The IPCC lists active travel and plant-based eating as "win wins" for health and climate change: IPCC, *Global Warming of 1.5°C: An IPCC Special Report on the Impacts of Global Warming of 1.5°C above Pre-Industrial Levels and Related Global Greenhouse Gas Emissions Pathways, in the Context of Strengthening the Global Response to the Threat of Climate Change, Sustainable Development, and Efforts to Eradicate Poverty* (2019), 714.

8. This is based on a UK study: Marko Tainio et al., "Can Air Pollution Negate the Health Benefits of Cycling and Walking?," *Preventive Medicine* 87 (2016)): 233–236.

9. The sources are Kirk Warren Brown and Tim Kasser, "Are Psychological and Ecological Well-Being Compatible? The Role of Values, Mindfulness, and Lifestyle," *Social Indicators Research* 74, no. 2 (2005): 349–368; Laura Fernanda Barrera-Hernández et al., "Connectedness to Nature: Its Impact on Sustainable Behaviors and Happiness in Children," *Frontiers in Psychology* 11, no. 276 (2020), https://doi.org/10.3389/fpsyg.2020.00276.

10. The Yale study is "Global Warming's Six Americas," Yale Program on Climate Change Communication, https://climatecommunication.yale.edu/about/projects/global-warmings-six-americas/ (accessed August 23, 2022). The figures are from September 2021. The Londoner's quote is from Tollemache, "We Have to Talk About . . . Climate Change," 231.

11. The Londoner's comment is from "We Have to Talk About . . . Climate Change," 232. The Stoknes quote is from *What We Think About When We Try Not to Think About Global Warming: Toward a New Psychology of Climate Action* (White River Junction, VT: Chelsea Green Publishing, 2015), 32.

12. The philosopher/psychologist study was published as J. Nagel and M. R. Waldmann, "Deconfounding Distance Effects in Judgments of Moral Obligation," *Journal of Experimental Psychology, Learning, Memory, and Cognition* 39, no. 1 (2013): 237–252.

13. Stoknes discusses cognitive dissonance in *What We Think About*, 67. My discussion of the smoking case is based on his analysis.

14. For a detailed and chilling exposé of climate denial, see Naomi Oreskes and Erik M. Conway, *Merchants of Doubt: How a Handful of Scientists Obscured the Truth on Issues from Tobacco Smoke to Global Warming* (New York: Bloomsbury, 2010). More recently, Geoffrey Supran and Naomi Oreskes revealed how ExxonMobil "systematically misled" the public on climate change: "Assessing ExxonMobil's Climate Change Communications (1977–2014)," *Environmental Research Letters* 12, no. 8 (2017), https://iopscience.iop.org/article/10.1088/1748-9326/aa815f; Supran and Oreskes, "Addendum to 'Assessing ExxonMobil's Climate Change Communications (1977–2014)," *Environmental Research Letters* 15, no. 11 (2020), https://iopscience.iop.org/article/10.1088/1748-9326/ab89d5/pdf. IPCC scientist Peter Stott brings the story up to date in *Hot Air: The Inside Story of the Battle against Climate Change Denial* (London: Atlantic Books, 2021). COVID vaccine refusers are less likely to get information from traditional sources

like TV, radio, government sources or newspapers, and more likely to seek it from social media, and to believe in conspiracy theories; Jamie Murphy et al., "Psychological Characteristics Associated with COVID-19 Vaccine Hesitancy and Resistance in Ireland and the United Kingdom," *Nature Communications* 12, no. 1 (2021): 2041–1723.

15. Norgaard's book is *Living in Denial: Climate Change, Emotions, and Everyday Life* (Cambridge, MA: MIT Press, 2011).

16. Hamilton's paper is "What History Can Teach Us About Climate Change Denial," in Weintrobe, *Engaging with Climate Change*, 16–32. For a brief account of the resistance (in the UK) and ridicule (in the United States) that met Einstein's breakthrough, see Matthew Wills, "Why No One Believed Einstein," *JSTOR Daily*, August 19, 2016. The climate change study is Aaron M. McCright and Riley E. Dunlap, "Cool Dudes: The Denial of Climate Change among Conservative White Males in the United States," *Global Environmental Change* 21, no. 4 (2011): 1163–1172. Hamilton makes the point about their cultural identity.

17. Michael Rustin makes the observation about consumption and success: "How Is Climate Change an Issue for Psychoanalysis," in Weintrobe, *Engaging with Climate Change*, 170–185. Weintrobe's argument is from "On the Love of Nature and on Human Nature: Restoring Split Internal Landscapes," *Engaging with Climate Change*, 200–213. All quotes from Andrews are from a Zoom interview with the author on September 17, 2021.

18. The term "modern neurosis" is from Johannes Lehtonen and Jukka Välimäki, "Discussion: The Difficult Problem of Anxiety in Thinking About Climate Change," in Weintrobe, *Engaging with Climate Change*, 48–51.

19. The study is Noah J. Goldstein, Robert B. Cialdini, and Vladas Griskevicius, "A Room with a Viewpoint: Using Social Norms to Motivate Environmental Conservation in Hotels," *Journal of Consumer Research* 35, no. 3 (2008): 472–482.

20. The Stoknes quote is from *What We Think About*, 96. Judith Lichtenberg explains this effect via the presence of other duty bearers, and embarrassment in *Distant Strangers*, 143–149.

21. The quote is from Dale Jamieson, "Climate Change, Responsibility, and Justice," *Science and Engineering Ethics* 16 (2010): 436. The point about

justifying and motivating reasons being different is from Stephen Gardiner, "Is No-One Responsible for Global Environmental Tragedy?"

22. The philosopher Simon Caney gives an excellent argument against discounting (which I have borrowed from here) in "Human Rights, Climate Change, and Discounting," *Environmental Politics* 17, no. 4 (2008): 536–555.

CHAPTER 9

1. The film is *Don't Look Up*, directed by Adam McKay (Netflix, 2021).

2. The film is *Independence Day*, directed by Roland Emmerich (20th Century Studios, 1996).

3. The Lichtenberg quote is from *Distant Strangers*, 232.

4. The letters are collected in "This Is How Scientists Feel," Is This How You Feel? website.

5. Charles's quote is from Tollemache, "We Have to Talk About . . . Climate Change," 221. The novel is Charles Dickens, *Great Expectations* (London: Chapman and Hall, 1861). For Randall's argument, see "Great Expectations: The Psychodynamics of Ecological Debt," in Weintrobe, *Engaging with Climate Change*, 90.

6. I owe these definitions (and the first quote) to Sally Weintrobe, *Engaging with Climate Change*, 6–8. On doublethink and perverse thinking, see Paul Hoggett, "Climate Change in a Perverse Culture," in Weintrobe, *Engaging with Climate Change*, 57–71; John Keene, "Unconscious Obstacles to Caring for the Planet," in Weintrobe, *Engaging with Climate Change*, 144–159.

7. Hoggett's argument is in *Climate Psychology*, 13. He channels the psychoanalyst Wilfried Bion. On suppressing emotion by blaming others, see Tollemache, "We Have to Talk About . . . Climate Change," 220–221. Nadine Andrews also highlighted the dangers of anxiety spiraling into climate fatalism in Zoom interview with the author, September 17, 2021. Heglar uses her inspired term in "Home Is Always Worth It," *Medium*, September 12, 2019.

8. The Yale study is "Global Warming's Six Americas," also discussed in chapter 8.

9. The quote is from Weintrobe, *Engaging with Climate Change*, 11.

10. The Stoknes quotes are from *What We Think About*, 171.

11. Greta Thunberg made this comment in a speech to the 2019 Davos Economic Forum; see Thunberg, *No-One Is Too Small to Make a Difference* (London: Penguin, 2019), 24. The quotes are from Hoggett, "Climate Change in a Perverse Culture," 85; Stoknes, *What We Think About*, 222.

12. For the full scientists' letters, see "This Is How Scientists Feel," Is This How You Feel? website.

13. Robinson's utopian novels include *Pacific Edge* (London: Unwin Hymàn, 1990), part of the *Three Californias* trilogy, and *The Ministry for the Future* (London: Orbit, 2020). The psychologists are Lehtonen and Välimäki, "Discussion," 48.

14. The Stoknes quote is from *What We Think About*, 67.

15. According to climate psychologist Rosie Robison, we can't expect perfect consensus at the start. Instead, the process of working together involves stepping outside one's own defensive barriers, building empathy despite stress, keeping going even without acknowledgment. It's about building visions, making countless small decisions and changes, empathizing with one another, acknowledging each other, seeking support, recognizing excuses in ourselves, as well as each other. And *keeping going*. See "Emotional Work as a Necessity: A Psychosocial Analysis of Low-Carbon Energy Collaboration Stories," in Hoggett, *Climate Psychology*, 85–106.

16. The DearTomorrow website is https://www.deartomorrow.org/. The quotes that follow are taken from it.

17. For more on the "Seven Generations" idea, see "Constitution of the Iroquois Nations," https://www.indigenouspeople.net/iroqcon.htm.

18. The Read quote is from *Parents for a Future*, 41.

19. The information on Maiden Castle is from "History of Maiden Castle," English Heritage, https://www.english-heritage.org.uk/visit/places/maiden-castle/history/.

20. Bjornerud's book is *Timefulness: How Thinking Like a Geologist Can Help Save the World* (Princeton: Princeton University Press, 2018). The "Anthropocene" might be better called the "Capitalocene" or "White (M)Anthropocene"; Jason Moore, *Anthropocene or Capitalocene? Nature, History, and the Crisis of Capitalism* (Oakland: PM Press, 2016); Giovanna Di Chiro,

"Welcome to the White (M)Anthropocene?," in *Routledge Handbook of Gender and Environment*, ed. Sherilyn MacGregor (Oxford: Routledge, 2017), 487–505.

21. Ialenti borrows from the strategies of so-called Safety Case experts, helping determine Finland's nuclear waste disposal policy; Ialenti, *Deep Time Reckoning: How Future Thinking Can Help Earth Now* (Cambridge, MA: MIT Press, 2020). Farrier's book is *Footprints: In Search of Future Fossils* (London: HarperCollins, 2020). For a visual projection of sea level rises, see Climate Central, "Coastal Risk Screening Tool: Land Projected to Be Below Annual Flood Level in 2050," https://sealevel.climatecentral.org/maps/ (accessed November 10, 2022).

22. The quote is from Bjornerud, *Timefulness*, 125.

23. For more on the limits of empathy, see Ray, *A Field Guide to Climate Anxiety*, 106–109.

24. The Jamieson quote is from "When Utilitarians Should Be Virtue Theorists," 182.

25. William Henry Davies, the Welsh poet who wrote those words, spent much of his unconventional life as a self-described "super-tramp" in the UK and the United States. See W. H. Davies, *Songs of Joy and Others* (London: Bibliolife, 1911).

26. On the rise in urban birdwatching, see Jacey Fortin, "The Birds Are Not on Lockdown, and More People Are Watching Them," *New York Times*, May 29, 2020.

27. See Randall and Hoggett, "Engaging with Climate Change."

Further Reading

(Some of these are academic texts, but some are written for a general audience. I've marked these with an asterisk, to make life easier.)

ON MORAL DUTIES: INDIVIDUAL AND COLLECTIVE

Cripps, Elizabeth. *Climate Change and the Moral Agent: Individual Duties in an Interdependent World*. Oxford: Oxford University Press, 2013.

Gardiner, Stephen M. *A Perfect Moral Storm: The Ethical Tragedy of Climate Change*. Oxford and New York: Oxford University Press, 2011.

Jamieson, Dale. "Climate Change, Responsibility, and Justice." *Science and Engineering Ethics* 16 (2010): 431–445.

Lichtenberg, Judith. *Distant Strangers: Ethics, Psychology, and Global Poverty*. New York: Cambridge University Press, 2014.

Mill, John Stuart. "On Liberty." In *John Stuart Mill: On Liberty and Other Essays*, ed. J. Gray (Oxford: Oxford University Press, 1991), 1859.

Singer, Peter. "Famine, Affluence, and Morality." *Philosophy and Public Affairs* 72, no. 1 (1972): 229–243.

ON THE ETHICS OF PARENTING

Archard, David, and David Benetar. *Procreation and Parenthood: The Ethics of Bearing and Rearing Children*. Oxford: Oxford University Press, 2010.

Brighouse, Harry, and Adam Swift. *Family Values: The Ethics of Parent-Child Relationships*. Princeton: Princeton University Press, 2014.

Cripps, Elizabeth. "Justice, Integrity, and Moral Community: Do Parents Owe It to Their Children to Bring Them up as Good Global Climate Citizens?" *Proceedings of the Aristotelian Society* 117, no. 1 (2017): 41–59.

Cripps, Elizabeth. "Do Parents Have a Special Duty to Mitigate Climate Change?" *Politics, Philosophy & Economics* 16, no. 3 (2017): 308–325.

Clayton, Matthew. *Justice and Legitimacy in Upbringing*. Oxford: Oxford University Press, 2006.

Fowler, Tim. *Liberalism, Childhood and Justice: Ethical Issues in Upbringing*. Bristol, UK: Bristol University Press, 2021.

*Swift, Adam. *How Not to Be a Hypocrite: School Choice for the Morally Perplexed Parent*. London and New York: Routledge, 2003.

ON GLOBAL EMERGENCIES

*Cripps, Elizabeth. *What Climate Justice Means and Why We Should Care*. London: Bloomsbury, 2022.

Intergovernmental Panel on Climate Change. *Sixth Assessment Reports*. https://www.ipcc.ch/report/sixth-assessment-report-cycle/.

*McGhee, Heather. *The Sum of Us: What Racism Costs Everyone and How We Can Prosper Together*. London: Profile Books, 2021.

*Oreskes, Naomi, and Erik M. Conway. *Merchants of Doubt: How a Handful of Scientists Obscured the Truth on Issues from Tobacco Smoke to Global Warming*. New York: Bloomsbury, 2010.

Sriram, Aditi, et al. *The State of the World's Antibiotics 2021: A Global Analysis of Antimicrobial Resistance and Its Drivers*. Washington, DC: Center for Disease Dynamics, Economics and Policy, 2021.

*Thunberg, Greta. *The Climate Book*. London: Allan Lane, 2022.

ON MORALITY AND HAPPINESS

*Hurka, Thomas. *The Best Things in Life: A Guide to What Really Matters*. Oxford and New York: Oxford University Press, 2011.

Mulnix, Jennifer Wilson, and M. J. Mulnix. *Happy Lives, Good Lives: A Philosophical Examination*. Ontario: Broadview Press, 2015.

Nussbaum, Martha. *Women and Human Development: The Capabilities Approach*. Cambridge: Cambridge University Press, 2000.

ON ACTIVISM AND ALLYSHIP

Chenoweth, Erica, and Maria J. Stephan. *Why Civil Resistance Works: The Strategic Logic of Nonviolent Conflict*. New York: Columbia University Press, 2012.

*Hayhoe, Katharine. *Saving Us: A Climate Scientist's Case for Hope and Healing in a Divided World*. Atria: One Signal Publishers, 2021.

*Kendi, Ibram X. *How to Be an Antiracist*. New York: Vintage, 2019.

*Nakate, Vanessa. *A Bigger Picture: My Fight to Bring a New African Voice to the Climate Crisis*. London: Pan Macmillan.

Nardini, Gia, Tracy Rank-Christman, Melissa G. Bublitz, Samantha N. N. Cross, and Laura A. Peracchio. "Together We Rise: How Social Movements Succeed." *Journal of Consumer Psychology* 31, October 20, 2020, https://doi.org/10.1002/jcpy.1201.

*Nash, Meredith, Robyn Moore, Ruby Grant, and Tania Winzenberg. "'It's Not about You': How to Be a Male Ally." *The Conversation*, April 5, 2021, https://theconversation.com/its-not-about-you-how-to-be-a-male-ally-158134.

*Thunberg, Greta. *No One Is Too Small to Make a Difference*. London: Penguin, 2019.

ON RAISING GLOBAL CITIZENS

*Agarwal, Pragya. *Wish We Knew What to Say: Talking with Children About Race*. London: Dialogue Books, 2020.

*Harvey, Jennifer. *Raising White Kids: Bringing up Children in a Racially Unjust America*. Nashville: Abingdon Press, 2017.

*Moyer, Melinda Wenner. *How to Raise Kids Who Aren't Assholes: Science-Based Strategies for Better Parenting—from Tots to Teens*. London: Headline Home, 2021.

*Shugarman, Harriet. *How to Talk to Your Kids About Climate Change: Turning Angst into Action*. Gabriola Island, Canada: New Society Publishers, 2020.

*Wegner, Bobbi. *Raising Feminist Boys: How to Talk with Your Child About Gender, Consent and Empathy*. Oakland, CA: Harbinger Publications, 2021.

ON LIFESTYLE CHOICES

*Berners-Lee, Mike. *There Is No Planet B: A Handbook for the Make or Break Years*. Cambridge: Cambridge University Press, 2019.

Nefsky, Julia. "How You Can Help, without Making a Difference." *Philosophical Studies*, no. 174 (2017): 2743–2767.

Wynes, Seth, and Kimberly A. Nicholas. "The Climate Mitigation Gap: Education and Government Recommendations Miss the Most Effective Individual Actions." *Environmental Research Letters* 12 (2017), https://doi.org/10.1088/1748-9326/aa7541.

ON HAVING KIDS (OR NOT HAVING THEM)

Conly, Sarah. *One Child: Do We Have a Right to More?* New York: Oxford University Press, 2015.

*McKibben, Bill. *Maybe One: A Case for Smaller Families*. New York: Plume, 1998.

Mullin, Amy. *Reconceiving Pregnancy and Childcare: Ethics, Experience, and Reproductive Labor*. Cambridge Studies in Philosophy and Public Policy. Cambridge: Cambridge University Press, 2005.

Rieder, Travis. *Toward a Small Family Ethic: How Overpopulation and Climate Change Are Affecting the Morality of Procreation*. Cham, Switzerland: Springer, 2016.

Sasser, Jade S. *On Infertile Ground: Population Control and Women's Rights in the Era of Climate Change*. New York: NYU Press, 2018.

ON PSYCHOLOGY

Hoggett, Paul. *Climate Psychology: On Indifference to Disaster*. Cham, Switzerland: Palgrave Macmillan, 2019.

Norgaard, Kari Marie. *Living in Denial: Climate Change, Emotions, and Everyday Life*. Cambridge, MA: MIT Press, 2011.

*Ray, Sarah Jacquette. *A Field Guide to Climate Anxiety: How to Keep Your Cool on a Warming Planet.* Oakland: University of California Press, 2020.

*Stoknes, Per Espen. *What We Think About When We Try Not to Think About Global Warming: Toward a New Psychology of Climate Action.* White River Junction, VT: Chelsea Green Publishing, 2015.

Weintrobe, Sally. *Engaging with Climate Change: Psychoanalytic and Interdisciplinary Perspectives.* London and New York: Routledge, 2013.

Index